Theory of Atomic Nuclei

The dimension of a typical atom is 100.000 times larger than the size of its nucleus, however, around 99.975% of the mass of an atom is concentrated in its nucleus. This is why nuclear theory is sometimes called the theory of everything. **Theory of Atomic Nuclei: Foundations** selects some fundamental problems of the theory of atomic nuclei at low energy. It reviews specific nuclear physics issues in terms of historical development and evolution into more extended theories, which nowadays constitute complete parts of nuclear science. The book allows the reader to understand the structure of the nucleus, i.e., the mechanism that rules in this subatomic world. The book is intended for graduate physics students at universities and nuclear sciences and engineering students at technical universities. It can also be helpful for PhD students and junior researchers in nuclear physics. The book is based on lectures on the theory of atomic nuclei given by the authors at the Maria Curie Skłodowska University in Lublin and at doctoral schools in Poland, France, Germany, the Czech Republic, and China in recent years. It is based mostly on original scientific papers and extensive monographs on nuclear theory.

Key Features:

- Written as a compact introduction to theories of the atomic nucleus
- Despite its compact form, the book acquaints the reader with the basic nuclear models and helps them better understand them
- Provides the reader with a foundation to go on and navigate contemporary theories or models of the atomic nucleus independently and in more detail

Bożena Nerlo-Pomorska is a theoretical physicist long year academic teacher specializing in the theory of atomic nuclei. She graduated in physics from Maria Curie-Skłdowska University. She completed her doctoral studies at the Institute of Nuclear Research. After her doctorate, she was employed at the MCS University, initially as an assistant professor and then a full professor. She completed several extended foreign sabbaticals at universities in Hanover, Regensburg in Germany, and the University of Strasbourg in France. She was also a visiting professor at the Technical University of Munich. She is the author of one textbook for students and about 140 scientific articles, cited over 2,000 times in the world literature.

Krzysztof Pomorski is an expert in the theory of atomic nuclei and nuclear reactions. He graduated in theoretical physics from Maria Curie-Skłdowska University in Lublin, Poland. He defended his doctoral thesis at the Institute of Nuclear Research. He was then employed at his alma mater, where he became a full professor. At present he is employed at the National Centre for Nuclear Research in Warsaw. He completed several years of research internships at the Technical University of Munich, the University of Regensburg in Germany, and the University of Strasbourg in France. He was also a visiting professor at Tsinghua University in Beijing, China. He is the author of three academic textbooks and over 200 scientific articles, which have been cited about 5,000 times. In 2011, he received the Marian Smoluchowski Medal, the highest distinction awarded by the Polish Physical Society.

Theory of Atomic Nuclei
Foundations

Bożena Nerlo-Pomorska
Krzysztof Roman Pomorski

CRC Press
Taylor & Francis Group
Boca Raton London New York

CRC Press is an imprint of the
Taylor & Francis Group, an informa business

Designed cover image: National Centre for Nuclear Research

First edition published 2026
by CRC Press
2385 NW Executive Center Drive, Suite 320, Boca Raton FL 33431

and by CRC Press
4 Park Square, Milton Park, Abingdon, Oxon, OX14 4RN

CRC Press is an imprint of Taylor & Francis Group, LLC

ISBN: 978-1-041-07484-7 (hbk)
ISBN: 978-1-041-08003-9 (pbk)
ISBN: 978-1-003-64330-2 (ebk)

DOI: 10.1201/9781003643302

Typeset in Nimbus Roman font
by KnowledgeWorks Global Ltd.

Publisher's note: This book has been prepared from camera-ready copy provided by the authors.

To our Teachers

Contents

Preface

OVERVIEW

The book contains a selection of some fundamental problems of the theory of atomic nuclei at low energy. It begins with a review of specific nuclear physics issues regarding historical development and evolving into more extended theories, which nowadays constitute complete parts of nuclear science. It is based on lectures on the theory of atomic nuclei given by the authors at the Maria Curie Skłodowska University in Lublin, Poland, as well as at doctoral schools in Poland, France, Germany, Czech Republic, and China in recent years. The book is based mainly on original scientific papers and extensive monographs on nuclear theory. The bibliography list includes books with more details related to the topics discussed in this book. The book allows the reader to understand the structure of the nucleus, i.e., the mechanism that rules in this subatomic world. The dimension of a typical atom is 100,000 times larger than the size of its nucleus. On the other hand, around 99.975% of an atom's mass is concentrated in its nucleus. That is why the nuclear theory is sometimes called the 'theory of everything'. It proves that knowledge of nuclear structure is important, especially nowadays when nuclear power stations are becoming increasingly important in supplying us with clean energy.

AUDIENCE

The book is intended for graduate physics students at universities and nuclear sciences and engineering students at technical universities. It can also be helpful for PhD students and junior researchers in nuclear physics. Only some basic knowledge of higher mathematics and quantum mechanics is sufficient to follow the book's contents.

ACKNOWLEDGMENTS

We thank all the colleagues and collaborators who contributed to this book. It is impossible to mention all persons from physics centers in Lublin, Warsaw,

Munich, Regensburg, Strasbourg, and Beijing, with whom we discuss eral aspects of the models presented in this manual. We are incredibly g to our teachers, Professors: Adam Sobiczewski, Stanisław Szpikowsk Zdzisław Szymanski, who introduced us to this fascinating micro-w atomic nuclei. We also thank Ann Chapman, Danny Kielty, and Sanika editors of CRC-Press, for their help and many valuable comments.

ABOUT THE AUTHORS

Bożena Gabriela Nerlo-Pomorska is a theoretical physicist long ye demic teacher specializing in the theory of atomic nuclei. She gra in physics from Maria Curie-Skłodowska University in Lublin, Polan completed her doctoral studies at the Institute of Nuclear Research i saw. Her doctoral thesis on *Electric moments of multipole atomic nucl* carried out under the supervision of Professor Adam Sobiczewski. Af doctorate, she was employed at the MCS University, initially as an as professor and then a full professor. She completed several extended f sabbaticals at universities in Hanover, Regensburg in Germany, and th versity of Strasbourg in France. She was also a visiting professor at the nical University of Munich. She is the author of one textbook for stude about 140 scientific articles, cited over 2,000 times in the world literat

Krzysztof Roman Pomorski is an expert in the theory of atomic and nuclear reactions. He graduated in theoretical physics from Maria Skłodowska University in Lublin, Poland. He defended his doctoral th *Moments of inertia of atomic nuclei and their deformation dependenc* Institute of Nuclear Research in Warsaw under the supervision of Pr Adam Sobiczewski. He was then employed at his alma mater, where came a full professor. From 2005 to 2008, he was the Dean of the Fac Mathematics, Physics, and Computer Science at the Maria Curie-Skłod University. The National Centre for Nuclear Research in Warsaw cu employs him. He completed several years of research internships at the nical University of Munich, the University of Regensburg in Germar the University of Strasbourg in France. He was also a visiting profe Tsinghua University in Beijing, China. He is the author of three aca textbooks and over 200 scientific articles, which have been cited abou times. In 2011, he received the Marian Smoluchowski Medal, the high tinction awarded by the Polish Physical Society.

Both authors are a married couple, and they have got two children.

Introduction

The authors aim isto familiarize the reader with the essential areas of the atomic nucleus theory so that after reading this textbook, he can independently read more advanced books or original scientific articles in this field. This domain of science is usually divided into low-, medium– and high-energy nuclear physics. It is generally believed that at excitations of nuclei up to 15 MeV per nucleon, it remains only a system of interacting protons and neutrons. In nuclei with energy above this limit, i.e., energies measured already in GeV, nuclear reactions occur in which other elementary particles are involved, making their analysis more complex.

In recent times, elementary particle physics has been experiencing dynamic development in both experiment and theory, creating several new models up to the standard one based on relativistic, quantum field theory, four-dimensional space, and many basic objects such as quarks, gluons, leptons, bosons γ, W, Z, and Higgs particles. The Standard Model accurately describes the effects of electromagnetic and weak and strong interactions.

The theory of low-energy nuclei represents less spectacular but well-established research directions, such as decay processes, analysis of the properties of nuclei in the ground and excited states, nuclear fission, or heavy ion collisions. Nuclear theoreticians try to find formulas for effective nuclear interaction and construct phenomenological models by adjusting their parameters to the experimental data.

Efforts to explore the nature of nuclear forces give hope for an accurate solution to many-body problems in the nucleus or in the infinite nuclear matter. However, for heavier but finite systems, this problem becomes mathematically unsolvable. Thanks to Pauli's exclusion principle, many fermion properties can be accurately described by an average single-particle potential, even without proper knowledge of the nuclear interactions. This potential can be, e.g., obtained by the Hartree-Fock method by looking for self-consistent nuclear fields or by building phenomenological models. Although approximate, microscopic theories placing nucleons in potential wells of various shapes have found a broad spectrum of practical applications by giving many results consistent with experimental data and explaining basic phenomena in the atomic nuclei.

The shell model's huge advantages, especially after including of residual forces, give the mathematical solutions simplicity, conceptual transparency, classic analogies, and results close to experimental data. Microscopic theories based on the average potential and the residual interactions added to it describe many nuclei's static properties despite the phenomenological choice of parameters. They have also become a good starting point for analyzing collective excitation and nuclear dynamics.

The residual forces, not included in the average potential, may be diagonalized for light nuclei, e.g., by the Talmi and de-Shalit methods, or, following the Copenhagen School, divided into long and short-range forces and included, e.g., using a variation calculation. The long-range interaction is manifested through the change of nucleus shape, and it can be taken into account by adding deformation to the average potential. The short-range interaction can be approximated by the Dirac δ function or by the forces connecting nucleons into pairs (pairing). The effect of the short-range interaction is well described by the Bardeen-Cooper-Schrieffer (BCS) theory, borrowed by Bohr, Mottelson, and Pines from solid-state physics. The superconducting model of nuclei explains well the difference between properties of nuclei with even or odd numbers of one kind of nucleon based on this variational BCS theory.

Improved methods of calculating nuclei's potential energy (e.g., binding energy or mass) have joined the macroscopic liquid drop model with the shell and pairing effects. The shell effects in the nucleus are included mainly by the shell correction method. Based on the Fermi gas model and knowledge of the magic numbers in nuclei, Myers and Swiatecki. have formulated the first prescription for evaluating the shell correction energy. Strutinsky developed a more mathematically restricted method for more realistic average potentials. The main macroscopic properties of nuclei are described in the liquid-drop model or its extensions. The microscopic shell and pairing effects have subtle effects on the potential energy.

Knowledge of the relationship between the nucleus's potential energy and its deformation allows specifying the shape of the nucleus at a point corresponding to the minimum energy, i.e., the equilibrium deformation. Deformation significantly impacts the nucleus's energy and properties both in the ground and excited states. The analysis of collective motions like rotation, shape oscillations or fission of nuclei requires a knowledge of the potential energy surface and the mass parameters determining the collective properties of nuclei. Adding the collective degrees of freedom to the single-particle coordinates leads to a particular difficulty by an artificially increasing number of degrees of freedom. This problem is solved in the Generating Coordinate Method (GCM), which allows obtaining dynamic quantities characterizing a nucleus's collective properties in a proper mathematical way.

The diagram on the next page shows the content of our handbook for various branches of nuclear physics and related fields. This lecture covers material marked with arrows connecting various branches of nuclear physics, whose names are written in white boxes.

The frame layout of the book.

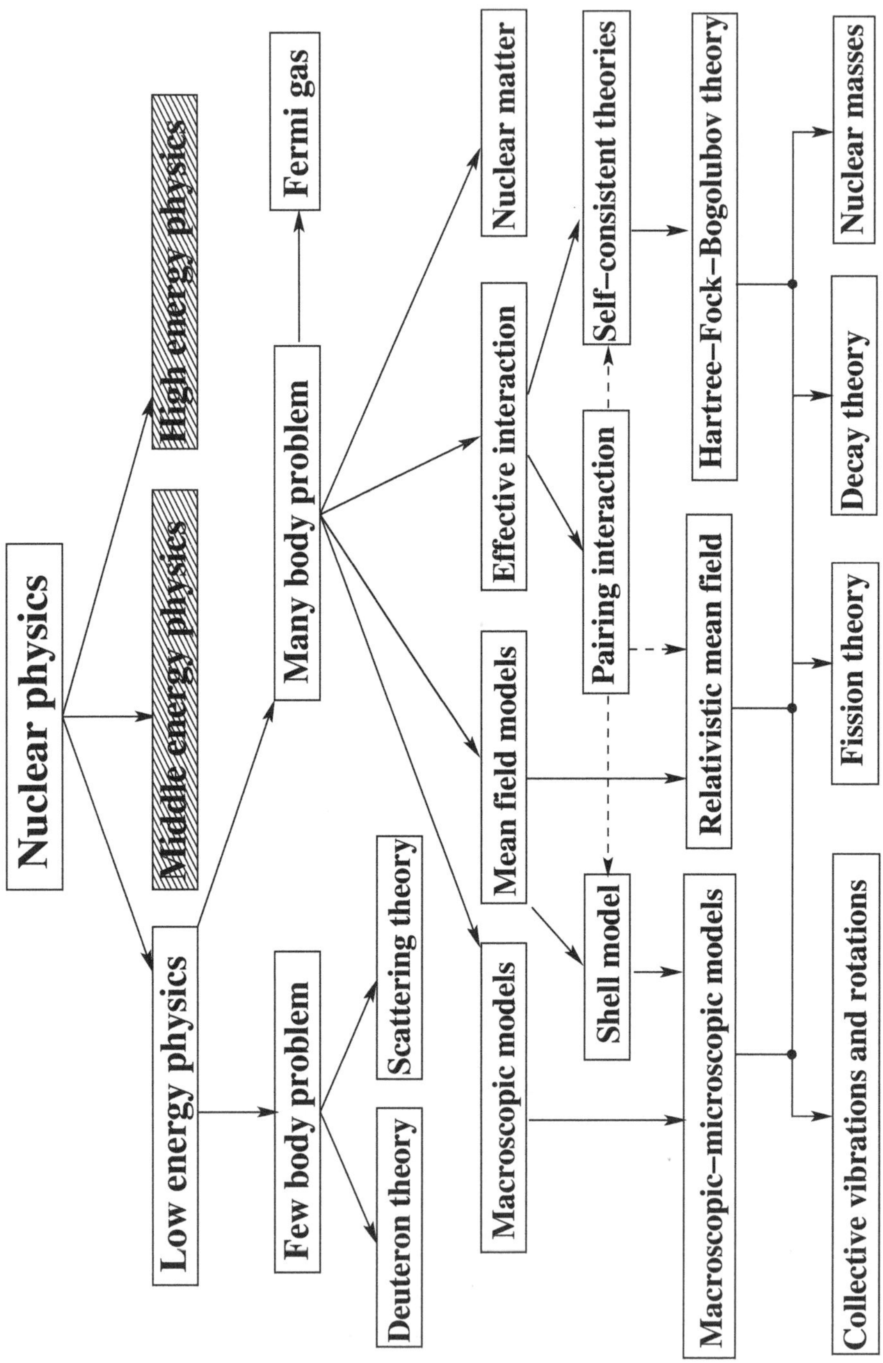

CHAPTER 1

Nuclear forces

1.1 INTERACTIONS IN NATURE

There are four types of fundamental interactions in nature corresponding to the observed forces: gravitational, electromagnetic, nuclear weak, and nuclear strong.

Gravity forces are responsible for phenomena in the macro world, e.g., they hold the planets in orbits. Newton's universal law describes them.

Electromagnetic forces acting between protons are 10^{37} times stronger than gravitational ones and counteract them for the same sign of charge. That is why there is no gravitational collapse. The competition between these two kinds of forces results in, e.g., finite density of planets. Electromagnetic interactions are responsible for the structure of atoms and molecules, determining their chemical properties. Coulomb's law and Maxwell's equations describe them, and the energy of these interactions in atoms is of the order of eV.

Weak nuclear forces are responsible for the β decay of atomic nuclei, that is, they are changing a neutron into a proton (or vice versa) with simultaneous electron (positron) and electron antineutrino (neutrino) emission. Both electromagnetic interactions and weak ones come from the electro-weak interactions.

Strong nuclear interactions , whose energy is measured in MeV, bind protons and neutrons into atomic nuclei and are responsible for the reactions between them. One of the theoretical nuclear physics objectives is to study these forces by interpreting the experimental facts using phenomenological models because there is no finite formula for a quantitative description of nuclear forces.

Micro-world phenomena are described using Schroedinger quantum mechanics with a potential that determines nuclear interactions. Information about such potential can be obtained by studying related double nucleon systems, as in deuteron, or unbound, as in scattering phenomena. Many properties of nuclear forces can be analyzed in the so-called nuclear matter model, built from an infinite number of nucleons. Initially, these considerations were an ‘academic’ theoretical model. However, the discovery of neutron stars has shown the full utility of this research. Nature can often surprise us. For nuclei built from a finite number of nucleons, it is impossible to solve many-body problems, even with knowledge of the exact form of interactions. Some phenomena, such as γ radiation and other types of radioactivity, dispersion, and cross-sections for reactions, can be described theoretically by passing many-body problems. Then, by building nuclear models, one simplifies, describing specific properties and omitting the others. We will focus on a non-relativistic theory sufficient to describe phenomena with energy below 15 MeV per nucleon. Of course, we appreciate efforts aimed at unifying interactions in nature by Einstein or Salam-Weinberg and attempts to explore the nature of nuclear forces through quantum chromodynamics useful in phenomenological nuclear models, in specific descriptions of the properties of nuclei from different mass regions.

1.2 WHICH MECHANICS DESCRIBES ATOMIC NUCLEUS

Newton’s classical mechanics is applied to describing phenomena in the macro-world when the wave nature of matter is not significant. In the micro-world, we are dealing with objects whose sizes are comparable with the corresponding lengths of the de Broglie waves.

Let us try to estimate the nucleon’s kinetic energy limit when Schroedinger’s quantum mechanics could be used. The typical size of nuclei (R_0) is of the order of fm (1 fm $=10^{-15}$ m), similar to the so-called classical radius of electron (R^e_{cl}), which is, according to the data in Table 1.1:

$$R_0 \approx R^e_{\mathrm{cl}} \equiv \frac{e^2}{m_e c^2} = 2.818\,\mathrm{fm}. \tag{1.1}$$

The Compton’s length of the de Broglie wave of a nucleus with M mass (Tables 1.1 and 1.2) is equal to:

$$\lambda = \frac{h}{Mc} = 1.32\,\mathrm{fm}. \tag{1.2}$$

Let us try to estimate what kinetic energy $T = \frac{p^2}{2M}$ has a nucleon with momentum p, whose wavelength λ_s is comparable with nucleus size, i.e. $\lambda_s = 2R_0$. By calculating the average kinetic energy of the nucleon in the nucleus from the formula

$$\lambda_s^2 = \left(\frac{h}{p}\right)^2 = \left(\frac{h}{Mc}\right)^2 \frac{(Mc)^2}{2TM} = \lambda^2 \frac{Mc^2}{2T},$$

we get the following equation:

$$T = \frac{Mc^2 \lambda_s^2}{8R_0^2} \cong 25\,\text{MeV}. \tag{1.3}$$

Table 1.1 Some physical constants

Name	Symbol	Value	Unit
Electron charge	e	$1.6022 \cdot 10^{-19}$	C
Electron charge square	e^2	1.44	MeV·fm
Mass of electron	m_e	$0.91094 \cdot 10^{-30}$	kg
Energy of electron	$m_e c^2$	0.51100	MeV
Speed of light in a vacuum	c	$2.9979 \cdot 10^{8}$	$\frac{m}{s}$
Planck's constant	h	$6.6261 \cdot 10^{-34}$	J·s
Dirac constant	$\hbar = \frac{h}{2\pi}$	$1.0546 \cdot 10^{-34}$	J·s
	$\hbar c$	197.3	MeV·fm
Fine structure constant	$e^2/\hbar c$	1/137.04	-
Nucleon mass	M	$1.674 \cdot 10^{-27}$	kg
Energy of nucleon	Mc^2	938.3	MeV
Nuclear magneton	$\mu_0 = \dfrac{e\hbar}{2Mc}$	$5.0508 \cdot 10^{-27}$	J/T
Dielectric constant of vacuum	ε_0	$8.8543 \cdot 10^{-12}$	F/m
Boltzmann constant	k	$1.3807 \cdot 10^{-23}$	J/K

The depth of the potential well created by the remaining particles must be slightly higher than the average kinetic energy of nucleons since nucleons are bound in the nucleus. Also, the nucleon's wave nature is important both outside and inside the nucleus, and quantum mechanics should be used. In general, we can limit ourselves to non-relativistic theory because the nucleon speed (v) in the nucleus is low compared to the speed of light c. Neglecting the relativistic effects, one makes an error of the order of the ratio value v^2/c^2, which is

$$\frac{v^2}{c^2} = \frac{2T}{Mc^2} \approx \frac{1}{20} = 5\%\,. \tag{1.4}$$

Because the uncertainties of the used models are, in many cases, more significant than relativistic corrections, one usually neglects them in describing low-energy phenomena and applies the Schroedinger equation, passing over the relativistic Dirac equation.

1.3 NUCLEAR PARTICLES

We assume that elementary particles involved in considered by us low-energy nuclear phenomena are nucleons. Their properties are given in Table 1.2.

Table 1.2 Properties of nucleons

Property	Symbol	Proton	Neutron
Energy	Mc^2	928.27231(28) MeV	939.56563(28) MeV
Spin	I	$\frac{1}{2}\hbar$	$\frac{1}{2}\hbar$
Life time	τ	$> 10^{26}$ years	889.1 ± 2.1 s
			$n \to p + e^- + \bar{\nu}_e$
Electric charge	q	$1 \pm 10^{-15}\,e$	$0 \pm 6 \cdot 10^{-12}\,e$
Magnetic moment	μ	2.7928474 μ_0	-1.9130428 μ_0

The concept of the atomic nucleus was introduced in 1911 by Rutherford, analyzing α particle scattering on thin gold films. He discovered that mass and positive charge are concentrated in the center of the atom. The presence of protons in the nucleus was discovered in dispersing high-energy particles α on nitrogen, where a faint glow was observed caused by protons extracted from nitrogen nuclei and obtained oxygen isotopes.

The neutron was discovered by Chadwick in 1930 by irradiating beryllium with α particles that generate a neutron flux and carbon. Neutrons are almost electrically neutral particles with a mass close to the mass of protons.

1.4 PROPERTIES OF NUCLEAR FORCES

Nuclear forces are:

a) attractive because they bind nucleons into nuclei though Coulomb repulsion,

b) repulsive at short distances because 'collapsing' of the nuclei is not observed; the nuclei have finite sizes,

c) short range, with a range of 10^{-15} m = 1 fm,

d) strong (energy of interactions, is calculated in MeV=10^6 eV),

e) static — mainly dependent on body distance $|\mathbf{r}_{12}|$: $V_{1,2} = V(|\mathbf{r}_{12}|)$,

f) independent of charge because forces acting between protons and neutrons are almost identical,

g) having saturation properties in nuclear matter.

Nuclear forces' statics and charge independence are not their fundamental properties because even interactions in a vacuum are dependent on the particles' angular momentum. Many phenomena indicate that forces between neutrons and protons and forces between acting particles of the same kind are not much different.

The dependence of the interaction potential of two nucleons from the distance of particles $r_{12} = |\mathbf{r}_1 - \mathbf{r}_2|$ is modeled in Fig. 1.1a. This interaction is made up of the short-range repulsive part (< 0.5 fm) – hardcore and the attractive Yukawa potential, disappearing exponentially with the increase of r_{12}. This potential is generated by the π meson field analogous to electromagnetic inter-atomic Van der Waals forces.

It should be noted that the interaction of two nucleons in a vacuum, described by the Yukawa function, or more realistic potentials: Hamada-Johnson, Paris, Argonne, or Bonn, differs from the effective two-body force in the nuclear matter because of the Pauli principle (applicable to protons and neutrons, which are fermions). Skyrme or Gogny potentials, with the parameters reproducing the properties of larger systems of nucleons, describe effective nuclear forces that act between two nucleons surrounded by nuclear matter. These potentials are used in the self-consistent method by Hartree and Fock to receive the mean-field potential of a multi-particle system.

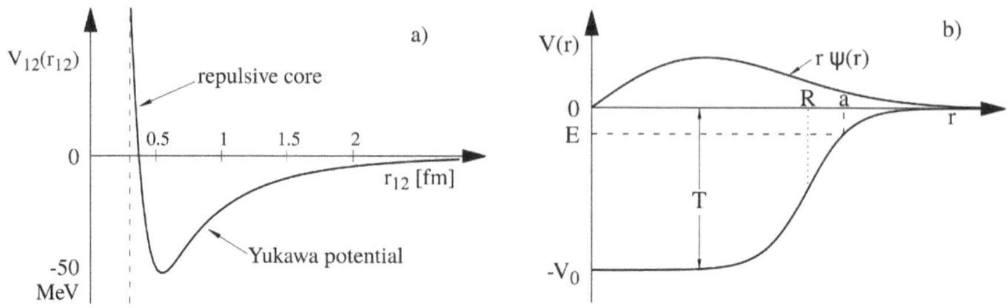

Figure 1.1 a) Dependence of the interaction potential of two nucleons in a vacuum V_{12} on the distance of particles r_{12}. b) Average nuclear potential $V(r)$ and radial part of the function $\psi(r)$ are denoted by solid lines, and r is the distance from the mass center of the nucleus.

Another more straightforward approach than the Hartree-Fock method is using phenomenological single-particle potential approximation. Such potential depends only on the distance r from the center of mass of the nucleus. A typical shape of the single-particle potential is shown in Fig. 1.1b. Since nuclear forces are short-range, they do not affect the properties of chemical molecules because they practically vanish outside the nucleus.

We must switch from laboratory to mass-center system to obtain the average potential between two nucleons. E.g. for a deuteron composed of a proton and a neutron, one considers the motion of a particle with a reduced mass of $M/2$ in the potential well $V(r)$. The potential range $\alpha = 2$ fm, slightly smaller than the area in which the particle is enclosed, is determined by the maximum probability of finding the particle at r, i.e., in the maximum of the function $r^2|\psi(r)|^2$. The radius R in Fig. 1.1b indicates the distance at which the probability $|\psi(R)|^2$ decreases e-times (here e=2.71828... is the natural logarithm base) compared to its value in the center $|\psi(0)|^2$. The function $\psi(r)$ is a radial part of the solution of the Schroedinger equation for the ground state. This problem will be discussed in more detail in Sec. 2.1.

For stable light nuclei, the number of Z protons and N neutrons are similar ($N \approx Z$). One has $N > Z$ in heavy nuclei, so the mass number equals $A = N + Z > 2Z$. Except for hydrogen, one does not observe nuclei consisting of only protons or neutrons, i.e., n-p interactions are necessary to form atomic nuclei.

The nuclear forces possess a saturation property, which courses the relationship of the experimental value of binding energy B of nucleons in the nucleus with the mass number A. The binding energy is defined as the amount of energy which one has to provide to the system to free its components (neutrons and protons), i.e., bring them far apart. This problem will be discussed in more detail in Chapter 5. It turns out that the binding energy per one nucleon is almost constant and equal to $B/A \approx 8$ MeV. In Fig. 1.2, it is visible that the nucleon binding energy in heavy nuclei is almost the same as in the α particle with three bonds per nucleon. It means that nucleons do not produce more than three bonds on the average. So, nuclear forces are saturated. As a result, the volume per nucleon in the nucleus is almost constant like it is in a drop of liquid. Therefore, the volume $\mathcal{V}$ of the entire nucleus is approximately proportional to its mass number $\mathcal{V} \sim A$. Hence, the conclusion is that the nucleus's radius is proportional to $A^{1/3}$: $R \sim A^{1/3}$.

The experimentally proved fact of saturation of nuclear force can be explained by the strong interaction of nucleons within the same orbital state $|n\,l\,m\rangle$, where n is main (energy), l – orbital, and m – magnetic quantum number. According to the Pauli-principle, such an orbital can only be occupied by four particles: two protons with opposite spins $2p(\uparrow\downarrow)$ and two neutrons $2n(\uparrow\downarrow)$. However, because the nucleus does not fall apart from α particles, there must be an interaction between different orbital states.

The saturation of nuclear forces is also justified by a strong repulsive core in the nucleon-nucleon interaction that does not allow nucleons to approach a small distance, limiting the number of neighbors to three. This happens likewise in liquids and solids. The property of saturation, common in liquid physics, has found justification in the theory of nuclei by analyzing the meson field and the internal structure of the nucleon. Apart from those considerations, we will limit ourselves to the properties of effective nuclear forces.

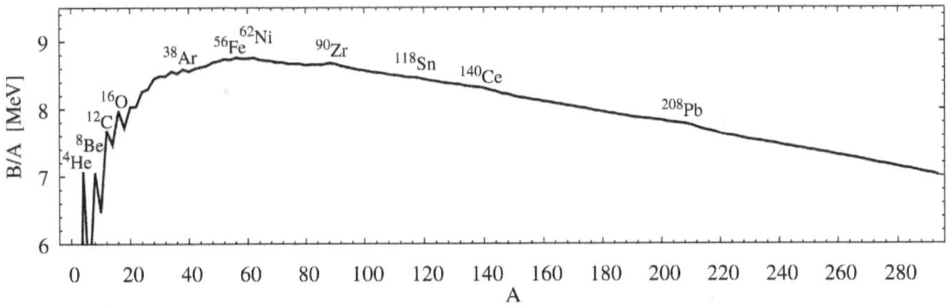

Figure 1.2 Nuclear binding energy per nucleon as a function of mass number A. The plot is made for the stable nuclei only.

Observing the curve in Fig. 1.2, one can see some deviations from the average value $B/A \approx 8$ MeV, which are no opposite to our conclusions. Slow fall for $A > 60$ occurs due to an increase of the Coulomb repulsion of protons with growing Z, while faster with A increase of binding energy in light nuclei are explained by the more significant relative role of the surface effects. For nuclei with proton numbers Z or neutrons N equal to 2, 8, 20, 28, 50, 82, and 126 for neutrons, one observes a rise in binding energy on the curve caused by shell effects. The numbers Z and N for which this effect appears are called magic. An increase in binding energy is considerable for doubly magic nuclei, i.e., having magic numbers of protons and neutrons. A closer analysis of energy relationships that enable the nuclear synthesis of lighter or fission of heavy nuclei will be carried out in Chapter 5.

Pauli's exclusion principle and the saturation property of nuclear forces allow describing many phenomena in heavy nuclei by constructing the average phenomenological potential, which represents the effective interaction of all nucleons on one chosen. The next chapter will analyze deuteron properties and nucleon scattering data and more detailed conclusions about the character of nucleon-nucleon interaction will be drawn there.

CHAPTER 2

Two body problem

2.1 DEUTERON THEORY

The deuteron is the most straightforward and only bound system of two nucleons: proton and neutron. Hence, it should provide us most information on the nature of nuclear forces. The theoretical models discussed in this chapter enable the description of primary experimental data for the deuteron. These data are collected in Table 2.1.

Table 2.1 Deuteron properties.

Name	Symbol	Value	Unit
Binding energy	B	2.226	MeV
Angular momentum	J	1	$\hbar$
Magnetic moment	μ	0.8573	μ_0
Electric quadrupole moment	$Q^{\exp}$	0.2859	$e\,\mathrm{fm}^2$
Charge radius	$R_{\mathrm{ch}}^{\exp}$	2.127	fm

2.1.1 Deuteron binding energy

Let us try to reproduce the numerical value of the ground state energy $E_{\mathrm{gs}} = -2.226$ MeV of the deuteron by a model which uses two parameters: central potential $V(r)$ defined by range $\alpha \approx R$ and depth V_0. Let us consider some of the simplest options here:

$$\text{potential of a rectangular well } V(r) = \begin{cases} -V_0 & \text{for } r \leq \alpha\,, \\ 0 & \text{for } r > \alpha\,, \end{cases} \tag{2.1}$$

$$\text{harmonic oscillator potential } V(r) = -V_0\left(1 - \frac{r^2}{2\alpha^2}\right), \tag{2.2}$$

$$\text{Gaussian potential } V(r) = -V_0\,\mathrm{e}^{-\frac{r^2}{\alpha^2}}\,, \tag{2.3}$$

$$\text{Fermi function potential } V(r) = \frac{-V_0}{1+\mathrm{e}^{\frac{r-R}{\alpha}}}\,, \tag{2.4}$$

All these functions approximate the shape of the realistic potential $V(r)$ shown in Fig. 1.1b. Wells with more complex shapes are closer to reality, but the analysis of low-energy phenomena shows that the model results are not very sensitive to the form $V(r)$. This is because the eigenfunction range is longer than the range of forces and is not affected by subtle effects inside the well. The potential of a rectangular form is best suited for an analytical solution. The aim is to find Hamilton's eigenfunction with this potential and choose V_0 and α parameters to reproduce the deuteron's experimental ground state energy E_{gs}.

Let us solve the Schroedinger equation for a particle in a rectangular potential well $V(r)$ with a depth of V_0 and α range as presented in Fig. 2.1. The problem of two nucleons of equal mass M in the center of mass coordinates will be solved, and therefore one considers the particle with a reduced mass $M_r = \frac{M \cdot M}{M+M} = M/2$. Such a particle is in the sphere with radius α because the whole problem has spherical symmetry, and one solves the Schroedinger equation in the radial coordinates.

In the case of the potential of a rectangular well, the Hamilton equation for a deuteron has the following form:

$$\left(\frac{-\hbar^2}{2M_r}\hat{\Delta} + V(r)\right)\psi = E\psi\,, \tag{2.5}$$

where the potential is defined as

$$V(r) = \begin{cases} -V_0\,, & r \leq \alpha \text{ in I}\,, \\ 0, & r > \alpha \text{ in II}\,, \end{cases} \tag{2.6}$$

and the wave function is a superposition of solutions in both partitions:

$$\psi(r) = \begin{cases} \psi_{I}(r) & \text{for } r \leq \alpha \\ \psi_{II}(r) & \text{for } r > \alpha\,. \end{cases} \tag{2.7}$$

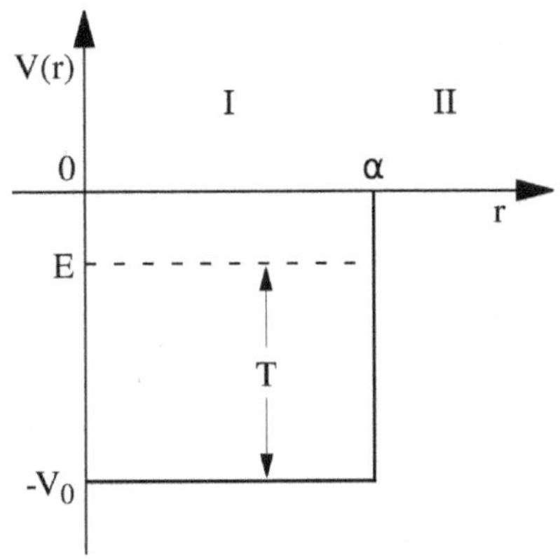

Figure 2.1 Rectangular potential well $V(r)$.

It should be remembered that in spherical coordinates, the Laplace operator $\hat{\Delta}$ consists of radial and angular parts

$$\hat{\Delta} = \frac{\partial^2}{\partial r^2} + \frac{2}{r}\frac{\partial}{\partial r} + \frac{1}{r^2}\hat{\Lambda}(\vartheta, \varphi), \tag{2.8}$$

where

$$\hat{\Lambda} = \cot\vartheta \frac{\partial}{\partial \vartheta} + \frac{\partial^2}{\partial \vartheta^2} + \frac{1}{\sin^2\vartheta}\frac{\partial^2}{\partial \varphi^2}.$$

So, for spherically symmetric potentials, the whole wave function can be presented as the product of $f(r)Y(\vartheta, \varphi)$. After extracting the radial and angular parts of the problem, one obtains a solution in the following form

$$\psi_{nlm}(r, \vartheta, \varphi) = f_{nl}(r)\, Y_{lm}(\vartheta, \varphi), \tag{2.9}$$

where the functions Y_{lm} are the spherical harmonics being the solution of the angular equation:

$$-\hat{\Lambda}(\vartheta, \varphi)Y_{lm}(\vartheta, \varphi) = l(l+1)Y_{lm}(\vartheta, \varphi), \tag{2.10}$$

where $l = 1, 2, 3, \ldots, m = -l, -l+1, \ldots, l$

At the same time, thanks to the proportionality of the angular momentum square operator to $\hat{\Lambda}$:

$$\hat{\mathbf{L}}^2 = -\hbar^2 \hat{\Lambda},$$

the Y_{lm} functions are eigenfunctions of $\hat{\mathbf{L}}^2$. The factor $l(l+1)$ plays the role of a constant to which one equalizes both sides of the Schroedinger equation, separated into an angular and radial part

$$\frac{r^2}{f}\left(\frac{\partial^2}{\partial r^2}+\frac{2}{r}\frac{\partial}{\partial r}\right)f+\frac{2M_{\mathrm{r}}(E-V(r))r^2}{\hbar^2}=-\frac{1}{Y}\hat{\Lambda}Y=l(l+1). \tag{2.11}$$

The radial equation takes the form:

$$\frac{\partial^2 f}{\partial r^2}+\frac{2}{r}\frac{\partial f}{\partial r}+\left(\frac{2M_{\mathrm{r}}(E-V(r))}{\hbar^2}-\frac{l(l+1)}{r^2}\right)f=0 \tag{2.12}$$

and for the constant potential $V(r) = V_0$ leads to the Bessel equation having solutions $f_{nl}(r)$ labeled by numbers

$$n = 1, 2, \ldots \text{ and } l = 0, 1, \ldots, n-1.$$

The ψ_{I} and ψ_{II} functions describing solution of the Schroedinger equation in both partitions of r must fulfill boundary conditions:

$$\begin{gathered}\psi_{\mathrm{I}}(\alpha)=\psi_{\mathrm{II}}(\alpha),\quad |\psi_{\mathrm{I}}(0)|\leq\infty,\\ \frac{\partial\psi_{\mathrm{I}}}{\partial r}|_{\alpha}=\frac{\partial\psi_{\mathrm{II}}}{\partial r}|_{\alpha},\quad \psi_{\mathrm{II}}(\infty)=0.\end{gathered} \tag{2.13}$$

The Schroedinger equation takes the form of a system of equations

$$\begin{cases}\left(-\frac{\hbar^2}{2M_{\mathrm{r}}}\hat{\Delta}-V_0\right)\psi_{\mathrm{I}}=E\psi_{\mathrm{I}},\\ \frac{-\hbar^2}{2M_r}\hat{\Delta}\psi_{\mathrm{II}}=E\psi_{\mathrm{II}}.\end{cases} \tag{2.14}$$

After dividing Eq. (2.14) by $\frac{-\hbar^2}{2M_{\mathrm{r}}}$ one gets

$$\begin{cases}\hat{\Delta}\psi_{\mathrm{I}}+\frac{2M_{\mathrm{r}}V_0}{\hbar^2}\psi_{\mathrm{I}}=\frac{-2M_{\mathrm{r}}E}{\hbar^2}\psi_{\mathrm{I}}\\ \hat{\Delta}\psi_{\mathrm{II}}=\frac{-2M_{\mathrm{r}}E}{\hbar^2}\psi_{\mathrm{II}}.\end{cases} \tag{2.15}$$

Let us introduce the notations:

$$K^2\equiv\frac{2M_{\mathrm{r}}V_0}{\hbar^2}=\frac{MV_0}{\hbar^2}>0, \tag{2.16}$$

$$k^2\equiv\frac{-2M_{\mathrm{r}}E}{\hbar^2}=\frac{-ME}{\hbar^2}>0. \tag{2.17}$$

Note that for the bound state, i.e., the state with $E < 0$, the parameter $k^2 > 0$. In this new notation, the system of equations (2.15) takes the form:

$$\begin{cases} \hat{\Delta}\psi_{\mathrm{I}} + (K^2 - k^2)\psi_{\mathrm{I}} = 0\,, \\ \hat{\Delta}\psi_{\mathrm{II}} - k^2\psi_{\mathrm{II}} = 0\,. \end{cases} \tag{2.18}$$

In the following, one assumes that

$$\kappa^2 \equiv K^2 - k^2 > 0\,, \quad \text{because} \quad V_0 > |E|\,. \tag{2.19}$$

This quantity is related to the kinetic energy of the particle in the potential well, since $V_0 - |E| = T$. Hence

$$\begin{cases} \hat{\Delta}\psi_{\mathrm{I}} + \kappa^2\psi_{\mathrm{I}} = 0\,, \\ \hat{\Delta}\psi_{\mathrm{II}} - k^2\psi_{\mathrm{II}} = 0\,. \end{cases} \tag{2.20}$$

where k is the wave vector. Let us limit ourselves to the solution corresponding to the ground state energy E_{gs}, i.e., the state with the lowest energy $E = E_{\mathrm{gs}}$ having quantum numbers $n = 1$, $l = 0$. Because $m = -l, -l+1, \ldots, l$, in our case $m = 0$, and the solution is only dependent on r

$$\psi(r,\vartheta,\varphi) = f_{10}(r)Y_{00} = \frac{1}{\sqrt{4\pi}} f_{10}(r)\,, \tag{2.21}$$

then one postulates

$$f_{10}(r) \equiv \frac{U(r)}{r}\,. \tag{2.22}$$

Then for energy $E = E_{\mathrm{gs}}$ and respectively $\kappa = \kappa_{\mathrm{gs}}$, $k = k_{\mathrm{gs}}$ system of equations (2.20) will take the form

$$\begin{cases} U_{\mathrm{I}}'' + \kappa_{\mathrm{gs}}^2 U_{\mathrm{I}} = 0\,, \\ U_{\mathrm{II}}'' - k_{\mathrm{gs}}^2 U_{\mathrm{II}} = 0\,, \end{cases} \tag{2.23}$$

with the following boundary conditions for the U function

$$\begin{aligned} U_{\mathrm{I}}(\alpha) &= U_{\mathrm{II}}(\alpha)\,, \\ U_{\mathrm{I}}'(\alpha) &= U_{\mathrm{II}}'(\alpha)\,. \end{aligned} \tag{2.24}$$

The functions $U_{\mathrm{I}}(r)$ and $U_{\mathrm{II}}(r)$ fulfill the following boundary conditions

$$U_{\mathrm{I}}(0) = 0\,, \quad \text{then} \quad \lim_{r\to 0} \frac{U_{\mathrm{I}}(r)}{r} = 1\,,$$

and

$$U_{\mathrm{II}}(\infty) = 0, \quad \text{then} \quad \lim_{r\to\infty} \frac{U_{\mathrm{II}}(r)}{r} = 0\,.$$

We are looking for solutions in the form of an exponential function because (2.23) is the harmonic oscillator equation.

$$U_{\mathrm{I}}(r) \sim \mathrm{e}^{\beta r}\,. \tag{2.25}$$

For the function U_{I} in area (I) one has $\beta^2 + \kappa_{\mathrm{gs}}^2 = 0$, hence $\beta = \pm\mathrm{i}\kappa_{\mathrm{gs}}$, so

$$U_{\mathrm{I}}(r) = A_1\mathrm{e}^{\mathrm{i}\kappa_{\mathrm{gs}}r} + A_2\mathrm{e}^{-\mathrm{i}\kappa_{\mathrm{gs}}r}\,. \tag{2.26}$$

From the condition $U_{\mathrm{I}}(0) = 0$, one gets $A_1 = -A_2$, so the function in the partition (I) takes the form

$$U_{\mathrm{I}} = A_1\left(\mathrm{e}^{\mathrm{i}\kappa_{\mathrm{gs}}r} - \mathrm{e}^{-\mathrm{i}\kappa_{\mathrm{gs}}r}\right) = 2A_1\mathrm{i}\,\sin(\kappa_{\mathrm{gs}}r) \equiv A_{\mathrm{gs}}\sin(\kappa_{\mathrm{gs}}r)\,. \tag{2.27}$$

Similarly for functions in the partition (II) $U_{\mathrm{II}}(r) \sim \mathrm{e}^{\gamma r}$ one has $\gamma^2 - k_{\mathrm{gs}}^2 = 0$, hence $\gamma = \pm k_{\mathrm{gs}}$, so

$$U_{\mathrm{II}}(r) = B_1\mathrm{e}^{k_{\mathrm{gs}}r} + B_{\mathrm{gs}}\mathrm{e}^{-k_{\mathrm{gs}}y}\,. \tag{2.28}$$

The condition $U_{\mathrm{II}}(\infty) = 0$ gives $B_1 = 0$ and it stays

$$U_{\mathrm{II}} = B_{\mathrm{gs}}\mathrm{e}^{-k_{\mathrm{gs}}r}\,. \tag{2.29}$$

So, one has a system of functions:

$$\begin{cases} U_{\mathrm{I}} = A_{\mathrm{gs}}\sin\kappa_{\mathrm{gs}}r & \text{for} \quad r \leq \alpha\,, \\ U_{\mathrm{II}} = B_{\mathrm{gs}}\,\mathrm{e}^{-\mathrm{i}k_{\mathrm{gs}}r} & \text{for} \quad r > \alpha\,. \end{cases} \tag{2.30}$$

Boundary conditions (2.24), often called stapling conditions, can be reduced to the equality of logarithmic derivatives of the functions of U at point α

$$\frac{U_{\mathrm{I}}'}{U_{\mathrm{I}}}|_\alpha = \frac{U_{\mathrm{II}}'}{U_{\mathrm{II}}}|_\alpha\,, \tag{2.31}$$

which gives the equation

$$\kappa_{\mathrm{gs}}\cot(\kappa_{\mathrm{gs}}\alpha) = -k_{\mathrm{gs}}\,. \tag{2.32}$$

The point k_{gs} satisfying this equation corresponds to the ground state energy

$$E_{\mathrm{gs}} = \frac{-\hbar^2\kappa_{\mathrm{gs}}^2}{M} - V_0\,. \tag{2.33}$$

Knowing the deuteron binding energy from measurement, one can estimate the parameters V_0 and α, which can reproduce this quantity in our model. Remembering that the kinetic energy $T \cong 25$ MeV and $E_{gs} = -2.226$ MeV, one can assume that $\kappa_{gs} \gg k_{gs}$, so $\kappa_{gs}^2 \approx K^2$ and $k_{gs}/\kappa_{gs} \approx 0$. Then the equation for eigenvalues takes the form $\cot(K\alpha) \approx 0$, so the lowest eigenvalue approximates the equation $K\alpha \approx \pi/2$ or $K^2\alpha^2 \approx \pi^2/4$. Hence, returning to the explicit form K^2 (2.16) one gets

$$\frac{M V_0 \alpha^2}{\hbar^2} = \frac{\pi^2}{4}. \tag{2.34}$$

The V_0 and α potential parameters are in the following relationship:

$$V_0 \alpha^2 = \frac{\pi^2 \hbar^2}{4M} \approx 103 \,\text{MeV fm}^2. \tag{2.35}$$

So, the allowed value of the ground-state binding energy of deuteron can be achieved by both deep and narrow rectangular well and shallow and wide ones.

To solve the eigenproblem of deuteron Hamiltonian, one still needs to find a connection between the constants A_{gs} and B_{gs}. One can use the normalization condition here

$$\int \psi^\star \psi \, d^3\mathbf{r} = 1. \tag{2.36}$$

This condition can be reduced to the product of the integral over angular coordinates and the sum of the radial integrals

$$\int_0^{2\pi} \int_{-1}^{1} Y^\star(\vartheta,\varphi)\, Y(\vartheta,\varphi)\, d\cos\vartheta d\varphi \left[\int_0^{\alpha} \frac{A_{gs}^2 \sin^2(\kappa_{gs} r)}{r^2} r^2 dr + \int_\alpha^\infty \frac{B_{gs}^2 \, e^{-2k_{gs} r}}{r^2} r^2 dr \right] = 1. \tag{2.37}$$

Since the spherical functions are normalized, it is sufficient to use the continuity condition (2.31) and the equation for eigenvalues (2.32), when calculating the radial integral. At α there is $\cot(\kappa_{gs}\alpha) = -k_{gs}/\kappa_{gs}$, so according to the relationship (2.19) one has $\sin(k_{gs}\alpha) = \kappa_{gs}/K$ and

$$A_{gs} \sin(\kappa_{gs}\alpha) = B_{gs}\, e^{-k_{gs}\alpha}, \tag{2.38}$$

or

$$B_{gs} = A_{gs} \frac{\kappa_{gs}}{K} e^{k_{gs}\alpha}. \tag{2.39}$$

The normalization condition (2.36) can therefore be rewritten in the form

$$A_{gs}^2\left\{\int_0^{\alpha}\sin^2(\kappa_{gs}r)dr+\frac{\kappa_{gs}^2}{K^2}\,e^{2k_{gs}\alpha}\int_{\alpha}^{\infty}e^{-2k_{gs}r}dr\right\}=1. \tag{2.40}$$

Substituting standard integral values

$$\int\sin^2(ax)\mathrm{d}x=\frac{1}{2}x-\frac{1}{4a}\sin(2ax) \tag{2.41}$$

and

$$\int e^{ax}\mathrm{d}x=\frac{1}{a}e^{ax}, \tag{2.42}$$

one obtains the constants in the form

$$A_{gs}=\left[\frac{2k_{gs}}{1+k_{gs}\alpha}\right]^{\frac{1}{2}},\quad B_{gs}=\frac{\kappa_{gs}}{K}\left[\frac{2k_{gs}}{1+k_{gs}\alpha}\right]^{\frac{1}{2}}e^{k_{gs}\alpha}. \tag{2.43}$$

Let us gather information obtained for the ground state of the deuteron. The wave function has the form

$$\begin{cases}\psi_{\mathrm{I}}=\dfrac{1}{\sqrt{4\pi}}\left[\dfrac{2k_{gs}}{1+k_{gs}\alpha}\right]^{\frac{1}{2}}\dfrac{\sin\kappa_{gs}r}{r} & for\ r\le\alpha\,,\\[2ex] \psi_{\mathrm{II}}=\dfrac{1}{\sqrt{4\pi}}\dfrac{\kappa_{gs}}{K}\left[\dfrac{2k_{gs}}{1+k_{gs}\alpha}\right]^{\frac{1}{2}}\dfrac{e^{k_{gs}(\alpha-r)}}{r} & for\ r>\alpha\,,\end{cases} \tag{2.44}$$

and the energy can be calculated from the relations (2.32) and (2.17)

$$\tan(\kappa_{gs}\alpha)=-\frac{\kappa_{gs}}{k_{gs}},\quad \text{where}\quad E_{gs}=\frac{-\hbar^2k_{gs}^2}{M}. \tag{2.45}$$

Let us solve Eq. (2.45) graphically as it is shown in Fig. 2.2. Plotting both sides of the equation as functions of $\kappa\alpha$, where $\kappa^2=K^2-k^2$, and $k^2=-ME/\hbar^2$, one is investigating the occurrence possibility of the intersection points of the function $\tan(\kappa\alpha)$ and $-\kappa\alpha/k\alpha$. One also assumes an unnaturally large value of K to discuss the solution more easily.

The examined ground state is the only bound state of the deuteron. To demonstrate this, one has to analyze the solution of Schroedinger equations, corresponding to the first excited state with energy E_{ex}, wave vector $k_{ex}^2=-ME_{ex}/\hbar^2$, $\kappa_{ex}^2=K^2-k_{ex}^2$, and $l=0$. In the second quadrant ($\pi/2<\kappa\alpha<\pi$) of Fig. 2.2 there is a solution for the first state E_{gs}. The next one should appear

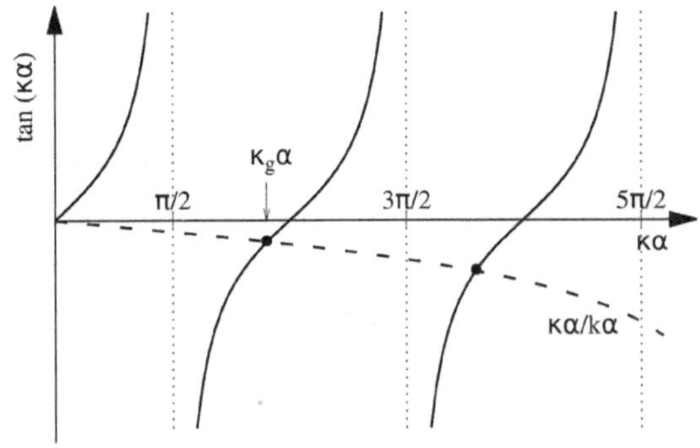

Figure 2.2 Graphical solution of the energy eigenproblem of deuteron with a rectangular potential well.

in the fourth quadrant. It will be shown that it is impossible for the range of nuclear forces $\alpha = 2$ fm. For the eigenstate in the second quarter, one has:

$$\frac{\pi}{2} < \kappa_{gs}\alpha < \pi\,, \quad \text{i.e.} \quad K^2\alpha^2 - k_{gs}^2\alpha^2 < \pi^2\,, \tag{2.46}$$

or otherwise

$$K^2\alpha^2 - \pi^2 < k_{gs}^2\alpha^2\,. \tag{2.47}$$

For the excited state in the fourth quarter, one would have:

$$\frac{3}{2}\pi < \kappa_{ex}\alpha < 2\pi\,. \tag{2.48}$$

Squaring this inequality and using the definition (2.19) one gets

$$K^2\alpha^2 - k_{ex}^2\alpha^2 > \frac{9}{4}\pi^2\,,$$

or otherwise

$$K^2\alpha^2 - \pi^2 > \frac{5}{4}\pi^2 + k_{ex}^2\alpha^2\,. \tag{2.49}$$

Comparing the inequalities (2.47) and (2.49), one gets:

$$\frac{5}{4}\pi^2 + k_{ex}^2 e\alpha^2 < K^2\alpha^2 - \pi^2 < k_{gs}^2\alpha^2\,. \tag{2.50}$$

If the excited state would be bound when: $E_{ex} < 0$ and $k_{ex}^2 > 0$. Then

$$k_{gs}^2\alpha^2 > \frac{5}{4}\pi^2\,, \tag{2.51}$$

hence

$$\alpha > \frac{\sqrt{5}}{2}\frac{\pi}{k_{gs}}, \tag{2.52}$$

which for the ground state energy of deuteron E_{gs} gives range 7 times greater than 2 fm. Therefore, the bound deuteron excited state with $l = 0$ cannot exist.

States with a higher angular momentum ($l > 0$) cannot fall below the examined E_{gs} because the energy increases with angular momentum in the considered example. For example, let us examine $l = 1$. The radial equation then takes the form:

$$\begin{cases} U_{\mathrm{I}}'' + \left(\kappa_{\mathrm{ex}}^2 - \dfrac{l(l+1)}{r^2}\right) U_{\mathrm{I}} = 0, \\ \\ U_{\mathrm{II}}'' + \left(k_{\mathrm{ex}}^2 - \dfrac{l(l+1)}{r^2}\right) U_{\mathrm{II}} = 0, \end{cases} \tag{2.53}$$

which gives for $l = 1$ solutions in the form of spherical and modified Bessel functions $j_1(\kappa_e r)$ and $\tilde{j}_1(k_e r)$:

$$\begin{cases} U_{\mathrm{I}} = A_{\mathrm{ex}}\left(\dfrac{\sin(\kappa_{\mathrm{ex}} r)}{\kappa_{\mathrm{ex}} r} - \cos(\kappa_{\mathrm{ex}} r)\right), \\ \\ U_{\mathrm{II}} = B_{\mathrm{ex}}\left(\dfrac{\mathrm{e}^{-k_{\mathrm{ex}} r}}{k_{\mathrm{ex}} r} + \mathrm{e}^{-k_{\mathrm{ex}} r}\right). \end{cases} \tag{2.54}$$

The boundary condition (2.31) of logarithmic derivatives of the function U_{I} and U_{II} at $r = \alpha$ gives the following equation for eigenvalues:

$$\kappa_{\mathrm{ex}}\alpha \cot(\kappa_{\mathrm{ex}}\alpha) = 1 + \left(\frac{\kappa_{\mathrm{ex}}}{k_{\mathrm{ex}}}\right)^2. \tag{2.55}$$

In the second quarter, the function $x\cot(x) < 1$, hence the left side of this equation is negative, while the right side remains positive. Thus, Eq. (2.55) may have a solution only in the third quarter, where $\cot(\kappa_{\mathrm{ex}}\alpha) > 0$. Then

$$\pi \le \kappa_{\mathrm{ex}}\alpha \le \frac{3}{2}\pi, \tag{2.56}$$

hence, considering the account (2.19), one gets an estimate

$$K^2\alpha^2 \ge \pi^2 + k_{\mathrm{ex}}^2\alpha^2. \tag{2.57}$$

This would lead to over twice larger value of the range α than was apparent from the discussions for the ground state (2.35). Hence

$$k_{\text{ex}}^2 \alpha^2 < 0 \quad \text{and} \quad E_{\text{ex}} > 0. \tag{2.58}$$

Therefore, the condition imposed on the ground state excludes a bound excited state of deuteron for $l = 1$. Numerical calculations for more realistic potentials than a rectangular well lead to similar conclusions.

The above considerations show that the nucleon-nucleon interaction depends on particles' orbital state l. Although in the ground state, nucleons interact the most in the same orbital state due to the saturation of nuclear forces. It is not the case for the nucleon-nucleon interaction in the excited states.

2.1.2 Angular momentum of deuteron

We will try to assemble the total momentum of deuteron $J = 1\hbar$ from spins of individual nucleons and orbital angular momentum L. Spins of each of two fermions are equal to $\frac{1}{2}\hbar$ and can be set in parallel to give an integer spin $S = 1\hbar$ or, on the contrary, giving $S = 0$. The total angular momentum of the system $J = L + S$ may correspond to the following orbital momentum (in the $\hbar$ units):

$$L = |J - S|, |J - S + 1|, \dots, |J + S|.$$

Table 2.2 contains all possible combinations of S and L for a deuteron having $J = 1$. The spectroscopic name of each state, generally accepted in nuclear physics, was also written there. Orbital angular momenta with $L = 0, 1, 2, \dots$ are marked accordingly by letters: S, P, D, ... The lower right index in symbols in the last row of the table means the total angular momentum and the left-upper one is the multiplicity of state: $2S + 1$.

Table 2.2 Angular momentum of a deuteron in $\hbar$ units.

J	1	1	1	1
S	0	1	1	1
L	1	0	1	2
${}^{2S+1}L_j$	${}^1\text{P}_1$	${}^3\text{S}_1$	${}^3\text{P}_1$	${}^3\text{D}_1$

In the previous section, one has shown that in the ground state of the deuteron orbital angular momentum $L = 0$, so this is the triplet state of $^3S_1$. That means that the state singlet $^1P_1$ would be higher in energy, which confirms the spin dependence of nuclear interaction.

2.1.3 Deuteron magnetic moment

Let us try to assemble the magnetic moment of the deuteron $\mu = 0.8573\mu_0$ from the proton $\mathbf{M}_p$ and the neutron $\mathbf{M}_n$ internal magnetic moments and the orbital magnetic moment originating from the proton. One can express the internal angular momenta through Pauli's spin matrices

$$\sigma_x = \begin{pmatrix} 0 & 1 \\ 1 & 0 \end{pmatrix}, \quad \sigma_y = \begin{pmatrix} 0 & -i \\ i & 0 \end{pmatrix}, \quad \sigma_z = \begin{pmatrix} 1 & 0 \\ 0 & -1 \end{pmatrix}, \tag{2.59}$$

and the numerical values of neutron μ_n and the proton μ_p magnetic moments taken from Table 1.2

$$\hat{\mathbf{M}}_p = \mu_p \hat{\sigma}_p, \quad \hat{\mathbf{M}}_n = \mu_n \hat{\sigma}_n. \tag{2.60}$$

We assume that half of the orbital moment of a deuteron originates from the angular proton momentum, and thus, the entire magnetic moment describes the operator

$$\hat{\mathbf{M}} = \mu_p \hat{\sigma}_p + \mu_n \hat{\sigma}_n + \frac{1}{2}\hat{\mathbf{L}}. \tag{2.61}$$

We used here the Bohr magneton units μ_0.

When determining the total spin $\hat{\mathbf{S}} = \frac{1}{2}(\hat{\sigma}_p + \hat{\sigma}_n)$ one can group the following terms in the operator $\hat{\mathbf{M}}$

$$\hat{\mathbf{M}} = \frac{1}{2}\left[\left(\mu_p + \mu_n + \frac{1}{2}\right)(\hat{\mathbf{S}} + \hat{\mathbf{L}}) + (\mu_p + \mu_n - \frac{1}{2})(\hat{\mathbf{S}} - \hat{\mathbf{L}}) + (\mu_p - \mu_n)(\hat{\sigma}_p - \hat{\sigma}_n)\right]. \tag{2.62}$$

This form will simplify calculating the average value $\hat{M}_z$ in a state with the maximum projection of $\mathbf{J}$ on the z-axis, i.e. value measured in the experiment. Let us construct the expectation value of the component $\hat{M}_z$ of the operator $\hat{\mathbf{M}}$ using total angular momentum $\mathbf{J}$, expressing the angles of projection of $\hat{\mathbf{M}}$ relative to the z-axis through the $\mathbf{J}$ components, as depicted in Fig. 2.3. Let us evaluate the component M_z of the magnetic moment with the vector $\mathbf{M}$ and the total angular momentum $\mathbf{J}$. Scalar product $\mathbf{M}\cdot\mathbf{J}$ is

$$\mathbf{M}\cdot\mathbf{J} = |\mathbf{M}||\mathbf{J}|\cos\alpha, \quad \text{i.e.} \quad \cos\alpha = \frac{\mathbf{M}\cdot\mathbf{J}}{|\mathbf{M}||\mathbf{J}|}, \tag{2.63}$$

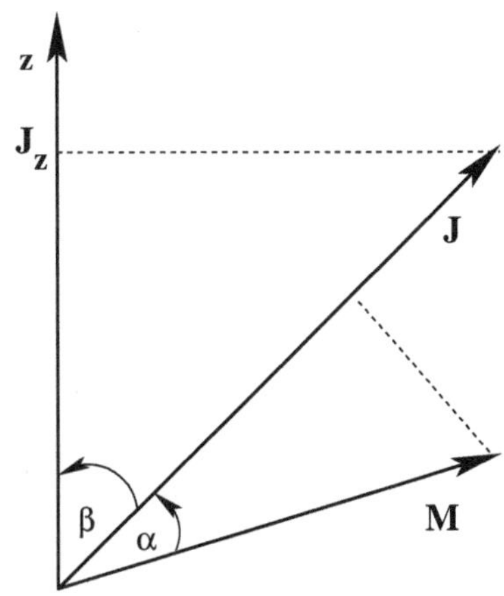

Figure 2.3 Angular momentum components.

where α is the angle between **M** and **J** as shown in Fig. 2.3.
The angular momentum component is

$$J_z = |\mathbf{J}|\cos\beta, \quad \text{i.e.} \quad \cos\beta = \frac{J_z}{|\mathbf{J}|}; \tag{2.64}$$

hence one can calculate the M_z magnetic moment component, which is precessing around **J**:

$$M_z = |\mathbf{M}|\cos\alpha\cos\beta = \frac{(\mathbf{M}\cdot\mathbf{J})J_z}{|\mathbf{J}||\mathbf{J}|}. \tag{2.65}$$

The last equality can be written for operators:

$$\hat{M}_z = \frac{(\hat{\mathbf{M}}\cdot\hat{\mathbf{J}})\hat{J}_z}{\hat{\mathbf{J}}^2}. \tag{2.66}$$

The average value of this operator in the state with the maximal projection of angular momentum $J_z = M = J$ is

$$\langle J|\hat{M}_z|J\rangle = \frac{\langle\hat{\mathbf{M}}\cdot\hat{\mathbf{J}}\rangle J}{(J+1)J}. \tag{2.67}$$

So, by inserting (2.62) in $\hat{\mathbf{M}}$ into the last formula, one obtains

$$\begin{aligned}\langle\hat{M}_z\rangle &= \frac{1}{2(J+1)}\Big[(\mu_\mathrm{p}+\mu_\mathrm{n}+\frac{1}{2})\langle\hat{\mathbf{J}}^2\rangle + (\mu_\mathrm{p}+\mu_\mathrm{n}-\frac{1}{2})\langle(\hat{\mathbf{S}}-\hat{\mathbf{L}})\cdot\hat{\mathbf{J}}\rangle + \\ &+ (\mu_\mathrm{p}-\mu_\mathrm{n})\langle(\hat{\sigma}_\mathrm{p}-\hat{\sigma}_\mathrm{n})\cdot\hat{\mathbf{J}}\rangle\Big].\end{aligned} \tag{2.68}$$

The operator $(\hat{\sigma}_\mathrm{p} - \hat{\sigma}_\mathrm{n})$ is antisymmetric with respect to the transposition of particles. Therefore it changes the symmetric triplet function with $S = 1$ (↑↑ , ↑↓, ↓↓) into an antisymmetric singlet with $S = 0$ (↑↓), so the average value of the last term in formula (2.68) disappears. Using dependencies $(\hat{\mathbf{S}} - \hat{\mathbf{L}}) \cdot (\hat{\mathbf{S}} + \hat{\mathbf{L}}) = \hat{\mathbf{S}}^2 - \hat{\mathbf{L}}^2$, one gets

$$\langle \hat{M}_z \rangle = \frac{J}{2}\left\{\left(\mu_\mathrm{p} + \mu_\mathrm{n} + \frac{1}{2}\right) + \left(\mu_\mathrm{p} + \mu_\mathrm{n} - \frac{1}{2}\right)\frac{S(S+1) - L(L+1)}{J(J+1)}\right\}. \tag{2.69}$$

Let us evaluate this quantity for the ground state and the other states, which according to Table 2.2 correspond to $J = 1\hbar$. Inserting into the formula (2.69) values of the proton and neutron magnetic moments (Table 1.2) and the quantum numbers from Table 2.2 one gets

$$\begin{aligned}
\langle \hat{M}_z \rangle_{3\mathrm{S}_1} &= \mu_\mathrm{p} + \mu_\mathrm{n} = 0.879\,, \\
\langle \hat{M}_z \rangle_{1\mathrm{P}_1} &= \frac{1}{2}\left[\left(\mu_\mathrm{p} + \mu_\mathrm{n} + \frac{1}{2}\right) - \left(\mu_\mathrm{p} - \mu_\mathrm{n} - \frac{1}{2}\right)\right] = 0.5\,, \\
\langle \hat{M}_z \rangle_{3\mathrm{P}_1} &= \frac{1}{2}\left(\mu_\mathrm{p} + \mu_\mathrm{n} + \frac{1}{2}\right) = 0.68\,, \\
\langle \hat{M}_z \rangle_{3\mathrm{D}_1} &= \frac{1}{2}\left(\mu_\mathrm{p} + \mu_\mathrm{n} + \frac{1}{2}\right) - \left(\mu_\mathrm{p} + \mu_\mathrm{n} - \frac{1}{2}\right) = 0.310\,.
\end{aligned} \tag{2.70}$$

As one can see, none of these numbers match the experimentally measured magnetic moment of deuteron $\mu = 0.8573\mu_0$ with a difference not exceeding the possible experimental error. It follows that the ground state of the deuteron is not a pure state with well-defined angular momentum described by the quantum number L. It means that the potential has no spherical symmetry, and the eigenfunction of Hamiltonian is a linear combination of states with different L. Starting from states with well-defined L, one can create all possible combinations from states with equal parity $(-1)^l$. The P states give the magnetic moment too small. Therefore, the ground state function of a deuteron must have the form:

$$\psi = a\psi_\mathrm{S} + b\psi_\mathrm{D}\,,$$

that is a mixed state of the triplet states: S and D.

Comparing the experimental value of the magnetic moment with the calculated one in this mixed state

$$\langle \hat{M}_z \rangle_\psi = |a^2| \langle \hat{M}_z \rangle_{3\mathrm{S}_1} + |b^2| \langle \hat{M}_z \rangle_{3\mathrm{D}_1} \tag{2.71}$$

and taking normalization of the wave function into account: $|a^2| + |b^2| = 1$, one will get $|b^2| = 0.039$. So about 4 percent of the admixture to the state $^3S_1$ is the contribution of the $^3D_1$ state.

When comparing the magnetic moment of a deuteron with the calculated moment in the pure L states, it should be stated that there is no reason for the magnetic moments in the nucleus not to be additive. Relativistic corrections are also in the order of 5%, and considering them would be significant. However, other deuteron properties, like the quadrupole moment, discussed in the next section, will also confirm the dependence of nuclear forces on angular momentum.

2.1.4 Electric quadrupole moment for a deuteron

Non zero deuteron quadrupole electric moment Q^{exp}=0.2859 $\text{e}\cdot\text{fm}^2$ indicates a spherically asymmetric charge distribution in the nucleus.

In spherical coordinates, the operator of the electric quadrupole moment has the following form:

$$\hat{Q} = e\sqrt{\frac{\pi}{5}}\, r^2 Y_{20}(\vartheta,\varphi)\,, \tag{2.72}$$

where Y_{20} is the spherical function whose explicit form is shown in Table 2.3. Other components $r^2 Y_{2m}(\vartheta,\varphi)$, which enter the quadrupole tensor, have the expected value equal to 0 in the state $|J = 1, M\rangle$.

The quadrupole moment Q measured in the experiment is equal to the mean value of the operator $\hat{Q}$ in the ground state with a maximum projection of the total angular momentum $M = J$

$$Q = \int \psi^*_{nJM}\, \hat{Q}\, \psi_{nJM}\, \mathrm{d}^3\mathbf{r}\,. \tag{2.73}$$

Suppose for a moment that the ground state wave function ψ_{nJM} is the product of the radial function $f_{nL}(r)$ and the angle function $Y_{LM_L}(\vartheta,\varphi)$ coupled with the spin function χ_{SM_s} to total angular momentum $J = 1$ with a third component $M = 1$

$$\psi_{nJM} = f_{nL}[Y_{LM_L}\chi_{SM_s}]_{JM}\,. \tag{2.74}$$

All such possible combinations of orbital and spin angles leading to $J = 1$ are listed in Table 2.2. The average values of $\hat{Q}$ in the states (2.74) are calculated using the angular momentum algebra rules and the properties of spherical functions. The matrix elements of the function Y_{20} and the corresponding Clebsch-Gordan coefficients can be found in textbooks, e.g., *Quantum Mechanics* by A.S. Davydov.

Table 2.3 Spherical harmonics $Y_{\lambda\mu}(\vartheta,\varphi)$ and Legendre polynomials $P_\lambda(\cos\vartheta) = \sqrt{\frac{4\pi}{2l+1}}\,Y_{\lambda 0}(\vartheta,0)$.

$Y_{00} = \sqrt{\frac{1}{4\pi}}$	$P_0(x) = 1$
$Y_{10} = \sqrt{\frac{3}{4\pi}}\cos\vartheta$	$P_1(x) = x$
$Y_{1\pm1} = \pm\sqrt{\frac{3}{8\pi}}\sin\vartheta e^{\pm i\varphi}$	
$Y_{20} = \sqrt{\frac{5}{16\pi}}(3\cos^2\vartheta - 1)$	$P_2(x) = \frac{1}{2}(3x^2 - 1)$
$Y_{2\pm1} = \pm\sqrt{\frac{15}{8\pi}}\sin\vartheta\cos\vartheta\,e^{\pm i\varphi}$	
$Y_{2\pm2} = \sqrt{\frac{15}{32\pi}}\sin^2\vartheta\,e^{\pm i\varphi}$	
$Y_{30} = \sqrt{\frac{7}{16\pi}}(5\cos^3\vartheta - 3\cos\vartheta)$	$P_3(x) = \frac{1}{2}(5x^3 - 3x)$
$Y_{3\pm1} = \mp\sqrt{\frac{21}{64}}(4\sin\vartheta\cos^2\vartheta - \sin^3\vartheta)\,e^{\pm i\varphi}$	
$Y_{3\pm2} = \sqrt{\frac{105}{32\pi}}\sin^2\vartheta\cos\vartheta e^{\pm 2i\varphi}$	
$Y_{3\pm3} = \sqrt{\frac{35}{64\pi}}\sin^2\vartheta\,e^{\pm 3i\varphi}$	
$Y_{40} = \sqrt{\frac{72}{16\pi}}(35\cos^4\vartheta - 30\cos^2\varphi + 3)$	$P_4(x) = \frac{1}{8}(35x^4 - 30x^2 + 3)$

The matrix element of the spherical harmonics $(Y_{\lambda\mu})$ between spherical eigenfunctions $|lm\rangle$ and $|l'm'\rangle$ is

$$\langle lm|Y_{\lambda\mu}|l'm'\rangle = \left(\frac{2\lambda+1}{4\pi}\right)^{-\frac{1}{2}}\left(\frac{2l+1}{2l'+1}\right)^{-\frac{1}{2}}(l'm'\lambda\mu|lm)(l'0\lambda 0|l0)\,, \qquad (2.75)$$

where $(l'm'\lambda\mu|lm)$ is the Clebsch-Gordan coefficient.
Hence, e.g.

$$\langle l=0, m=0|Y_{\lambda 0}|l=0, m=0\rangle = \frac{1}{\sqrt{4\pi}}\delta_{\lambda 0}\,. \tag{2.76}$$

Legendre polynomials meet the following orthogonality conditions:

$$\int_{-1}^{1} P_\lambda(x)P_{\lambda'}(x)\,\mathrm{d}x = \frac{2}{2\lambda+1}\delta_{\lambda\lambda'}\,. \tag{2.77}$$

There is also a relation:

$$P_\lambda(\cos(\vartheta_1-\vartheta_2)) = \frac{4\pi}{2\lambda+1}\sum_{\mu=-\lambda}^{\lambda} Y_{\lambda\mu}(\vartheta_1,\varphi_1)Y^\star_{\lambda\mu}(\vartheta_2,\varphi_2)\,. \tag{2.78}$$

Let us calculate, for example, the average value of the quadrupole moment operator $\hat{Q}$ at $^3\mathrm{S}_1$ at $J=1$, $M=1$, $L=0$, $S=1$:

$$\begin{aligned} Q_{^3S_1} &= e\sqrt{\frac{\pi}{5}}\langle n=1, L=0, M_L=0|r^2Y_{20}|n=1, L=0, M_L=0\rangle \\ &\quad \cdot \langle S=1, M_S=1|S=1, M_S=1\rangle \\ &= e\sqrt{\frac{\pi}{5}}\langle n=1, L=0|r^2|n=1, L=0\rangle \\ &\quad \cdot \langle L=0, M=0|Y_{20}|L=0, M=0\rangle = 0 \end{aligned} \tag{2.79}$$

At $^3\mathrm{S}_1$, a deuteron's nucleus would therefore be spherical.

Similarly, one calculates quadrupole moments for other states from Table 2.2. And so

$$Q_{^1\mathrm{P}_1} = -\frac{e}{10}\langle n=2, L=1|r^2|n=2, L=1\rangle < 0\,. \tag{2.80}$$

This result means that in the state $^1\mathrm{P}_1$, the nucleus would be oblate. Likewise, a simple calculation gives

$$Q_{^3\mathrm{P}_1} = \frac{e}{20}\langle n=2, L=1|r^2|n=2, L=1\rangle > 0\,. \tag{2.81}$$

It means that in the state $^3\mathrm{P}_1$ a deuteron nucleus would be prolate.

One can also show that, as the suggested in Sec. 2.1.3, by matching the magnetic moment mixing of S + D states

$$\psi = a\psi_\mathrm{S} + b\psi_\mathrm{D}\,, \tag{2.82}$$

has the quadrupole moment close to the experimental one. After relevant calculations carried out using the matrix elements (2.75) one obtains the quadrupole moment in the mixed state S + D :

$$Q_{\mathrm{S+D}} = \frac{8\pi e}{5}(a^* b u - b^2 w), \tag{2.83}$$

where

$$\begin{aligned} u &= \langle n = 1, L = 0 | r^2 | n = 3, L = 2 \rangle \\ w &= \langle n = 3, L = 2 | r^2 | n = 3, L = 2 \rangle \end{aligned} \tag{2.84}$$

are radial integrals of r^2 between the states S and D. After calculations we will get the result very close to the experimental value given in Table 2.1:

$$Q_{\mathrm{S+D}} = 0.27\, e \cdot \mathrm{fm}^2 \approx Q^{\mathrm{exp}}, \tag{2.85}$$

for the parameter $b = 0.04$.

2.1.5 Nuclear tensor force coupling

We are looking for such a neutron-proton interaction V, which is not central, causes the mixing of different orbital angular moment states, and distinguishes a certain direction in space. This interaction should meet the following conditions:

1. The dependence of the interaction on the direction can be determined via Pauli's spin operators $\hat{\sigma}_{\mathrm{n}}, \hat{\sigma}_{\mathrm{p}}$.

2. The angular dependence of the interaction should be proportional to the spherical function Y_{2m}, which gives a non-zero element matrix between states S and D

$$\int \psi_{00}\, V \psi_{2m}\, \mathrm{d}\Omega \neq 0\,.$$

3. The interaction should be scalar because energy is a scalar.

The operator which meets these conditions is

$$\hat{V} = V_{\mathrm{t}}(r)\, \hat{s}_{\mathrm{np}}, \tag{2.86}$$

where the scalar operator $\hat{s}_{\mathrm{np}}$ is built from tensors of the second rank

$$\hat{s}_{\mathrm{np}} = \sum_{\alpha\beta} T_{\alpha\beta} \hat{O}_{\alpha\beta}, \tag{2.87}$$

of which the first $T_{\alpha\beta}$ is a trace-less tensor in coordinate space, and its components are linear combinations of spherical functions Y_{2m}

$$T_{\alpha\beta} = \frac{3}{r^2}\left(x_\alpha x_\beta - \frac{1}{3}r^2\delta_{\alpha\beta}\right). \tag{2.88}$$

Here x_α are components of the $\mathbf{r}$ vector. The operator $\hat{O}_{\alpha\beta}$ is built with spin vectors

$$\hat{O}_{\alpha\beta} = (\hat{\sigma}_{\mathrm{n}})_\alpha(\hat{\sigma}_{\mathrm{p}})_\beta. \tag{2.89}$$

Finally, after some transformations one gets operator $\hat{s}_{\mathrm{np}}$ presented in the form

$$\hat{s}_{\mathrm{np}} = \frac{3(\hat{\sigma}_{\mathrm{n}}\cdot\mathbf{r})(\hat{\sigma}_{\mathrm{p}}\cdot\mathbf{r})}{r^2} - (\hat{\sigma}_{\mathrm{n}}\cdot\hat{\sigma}_{\mathrm{p}}). \tag{2.90}$$

The amplitude $V_{\mathrm{t}}(r)$ makes neutron-proton tensor interaction dependent on the distance of nucleons. The found interaction is not the only one possible, but it does have a property that disappears in the singlet state with the spin eequal to zero ($S = 0$). The tensor interaction V_{t} has to be added to the central part V_{c}

$$\hat{V}(r) = V_{\mathrm{c}}(r) + V_{\mathrm{t}}(r)\,\hat{s}_{\mathrm{np}}, \tag{2.91}$$

although one could also make the central interaction dependent on the state of the system (confer Sec. 2.1.6).

The central and tensor interaction is the Wigner potential. The Schroedinger equation with this potential leads to a coupled system of differential equations which can only be solved numerically

$$\left[\frac{-\hbar^2\hat{\Delta}}{M} + V_{\mathrm{c}}(r) + V_{\mathrm{t}}(r)\hat{s}_{\mathrm{np}}\right]\psi = E\psi. \tag{2.92}$$

2.1.6 General form of the nucleon-nucleon interaction

Properties of the saturation of nuclear forces cause interaction between nucleons in the same orbital state to be particularly strong. In the ground state of the deuteron, two nucleons occupy the same spatial state. However, this is not the case in the excited states. The dependence of potential on the state of the system is expressed using operators exchanging spatial coordinates, spin coordinates, or both.

Majorana operator $\hat{P}_{\mathrm{M}}$ exchanges the spatial coordinates of two particles. Let the function ψ describe the particle system:

$$\hat{P}_{\mathrm{M}}\psi(\mathbf{r}_1,\mathbf{r}_2) = \psi(\mathbf{r}_2,\mathbf{r}_1). \tag{2.93}$$

$\hat{P}_M$ does not change the state of the system if both particles (1 and 2) are in the same state.

If one divides the system state function into a symmetric part ψ^s and antisymmetric one ψ^a

$$\psi = \psi^s + \psi^a, \quad \text{where} \quad \hat{P}_M \psi^s = \psi^s, \ P_M \psi^a = -\psi^a, \tag{2.94}$$

so

$$\psi^s = \frac{1}{2}(\psi + \hat{P}_M \psi), \tag{2.95}$$

$$\psi^a = \frac{1}{2}(\psi - \hat{P}_M \psi) \tag{2.96}$$

and

$$\hat{P}_M \psi = \psi^s - \psi^a. \tag{2.97}$$

Majorana has constructed the potential in the form of:

$$\hat{V}_M = \frac{1}{2} \sum_{j,k,j \neq k} V_M(|\mathbf{r}_j - \mathbf{r}_k|)\hat{P}_M, \tag{2.98}$$

where V_M is a function of the usual attractive potential for particles in the same space state $\hat{V}_M = V_M$. In a system of particles in different orbital states, its impact is weakening, which leads to the saturation of nuclear forces.

Analogously, Bartlett introduced the exchange of spin coordinates of particles, and Heisenberg exchanged space and spin coordinates simultaneously. Let us gather information about nuclear interactions:

a) The potential of the nucleon-nucleon interaction depends on the following coordinates of particles 1 and 2 (see Fig. 2.4):
   - spatial coordinates: $\mathbf{r}_1$, $\mathbf{r}_2$,
   - spin: $\hat{\sigma}_1$, $\hat{\sigma}_2$,
   - isospin: $\hat{\tau}_1$, $\hat{\tau}_2$,

b) The central part of the potential is a function of particle distances $\mathbf{r}_{12} = \mathbf{r}_1 - \mathbf{r}_2$.

Let us also list the operators acting on the system wave function:

1. Wigner operator – no exchange term, no change of the wave function

$$\hat{P}_W \psi(\mathbf{r}_1, \hat{\sigma}_1; \mathbf{r}_2, \hat{\sigma}_2) = \psi(\mathbf{r}_1, \hat{\sigma}_1; \mathbf{r}_2, \hat{\sigma}_2). \tag{2.99}$$

2. Majorana operator that exchanges space coordinates

$$\hat{P}_M \psi(\mathbf{r}_1, \hat{\sigma}_1; \mathbf{r}_2, \hat{\sigma}_2) = \psi(\mathbf{r}_2, \hat{\sigma}_1; \mathbf{r}_1, \hat{\sigma}_2). \tag{2.100}$$

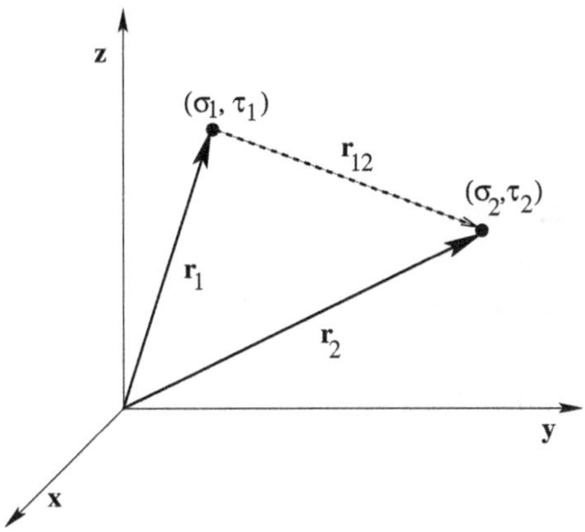

Figure 2.4 Coordinates of two nucleons in the nucleus.

3. Bartlett operator exchanging spin coordinates

$$\hat{P}_{\rm B}\psi(\mathbf{r}_1,\hat{\sigma}_1;\mathbf{r}_2,\hat{\sigma}_2) = \psi(\mathbf{r}_1,\hat{\sigma}_2;\mathbf{r}_2,\hat{\sigma}_1)\,. \tag{2.101}$$

The explicit form of this operator is $\hat{P}_{\rm B} = \frac{1}{2}(1+\hat{\sigma}_1\cdot\hat{\sigma}_2)$.

4. Heisenberg operator replacing spatial and spin coordinates

$$\hat{P}_{\rm H}\psi(\mathbf{r}_1,\hat{\sigma}_1;\mathbf{r}_2,\hat{\sigma}_2) = \psi(\mathbf{r}_2,\hat{\sigma}_2;\mathbf{r}_1,\hat{\sigma}_1)\,. \tag{2.102}$$

As one can see

$$(\hat{P}_{\rm W})^2 = (\hat{P}_{\rm M})^2 = (\hat{P}_{\rm B})^2 = (\hat{P}_{\rm H})^2 = 1 \tag{2.103}$$

and

$$\hat{P}_{\rm H} = \hat{P}_{\rm M}\hat{P}_{\rm B}\,. \tag{2.104}$$

The potential of nuclear forces can be represented as superposition of potentials corresponding to the Wigner's non-exchangeable force and the exchangeable forces of Majorana, Bartlett, and Heisenberg:

$$\begin{aligned}\hat{V} &= \frac{1}{2}\sum_{ij,i\neq j}\Big(V_c(\mathbf{r}_{ij}) + V_{\rm t}(\mathbf{r}_{ij})\hat{s}_{\rm np} + \\ &+ V_{\rm M}(\mathbf{r}_{ij})\,\hat{P}^{ij}_{\rm M} + V_{\rm B}(\mathbf{r}_{ij})\,\hat{P}^{ij}_{\rm B} + V_{\rm H}(\mathbf{r}_{ij})\,\hat{P}^{ij}_{\rm H}\Big)\,,\end{aligned} \tag{2.105}$$

where $\mathbf{r}_{ij} = \mathbf{r}_i - \mathbf{r}_j$.

There is no explicit isospin exchange operator in the potential (2.105)

$$\hat{P}_{\tau} = \frac{1}{2}(1 + \hat{\tau}_1 \cdot \hat{\tau}_2) \tag{2.106}$$

where $\hat{\tau}_1$ and $\hat{\tau}_2$ are the Pauli matrices corresponding to isospins. The following consideration proves that such an operator is no longer needed. When describing the neutron-proton system, one can treat both nucleons as identical particles with the only different isospin coordinates $\hat{\tau}_1, \hat{\tau}_2$. Then, the function of two fermions can be written as the product of the spatial, spin, and isospin parts

$$\Psi(\mathbf{r}_1, \hat{\sigma}_1, \hat{\tau}_1; \mathbf{r}_2, \hat{\sigma}_2, \hat{\tau}_2) = \psi(\mathbf{r}_1, \mathbf{r}_2) \cdot \chi(\hat{\sigma}_1, \hat{\sigma}_2) \cdot \eta(\hat{\tau}_1, \hat{\tau}_2). \tag{2.107}$$

Due to the Pauli's exclusion principle, this function is anti-symmetrical, i.e. must change the sign when all nucleons 1 and 2 coordinates are changed. Therefore, there must be equality

$$\hat{P}_{\mathrm{M}} \hat{P}_{\mathrm{B}} \hat{P}_{\tau} = -1\,. \tag{2.108}$$

Hence it follows that $\hat{P}_{\tau} = -\hat{P}_{\mathrm{H}}$. Therefore, there is no need for such an operator to appear explicitly in the potential nucleon-nucleon formula.

2.2 NEUTRON-PROTON SCATTERING

Neutron-proton scattering is tested by a neutron projectile bombarding a target containing hydrogen. Let us consider the neutron-proton system's unbound states, assuming that the neutron energy is small and does not exceed 5 MeV. So, it will be elastic scattering, where there is no energy loss due to the excitation of particles.

Let us calculate the mass-center ($\mathcal{MC}$) coordinates, assuming that the nucleon masses are equal. One then considers motion a fictitious particle with a mass of $M/2$ in a field with a $V(r)$ potential. The goal is to evaluate the scattering angle ϑ and the scattering cross-sections related to the wave function of the scattered particle. Fig. 2.5 shows the neutron-proton scattering in the laboratory system ($\mathcal{L}$). Table 2.4 gives the relationship between the quantities specified in the $\mathcal{MC}$ and the $\mathcal{L}$ systems.

In the laboratory system, proton and neutron, particles of equal mass M, run apart after an elastic collision in directions perpendicular to each other $\vartheta + \vartheta' = \pi/2$. In the mass-center coordinate system, the particle with mass $M/2$ disperses on the potential $V(r)$. Analyzing measurement results and neutrons scattered far from the mass center, we can limit ourselves to the theoretical analysis of asymptotic wave functions, shown in Fig. 2.6.

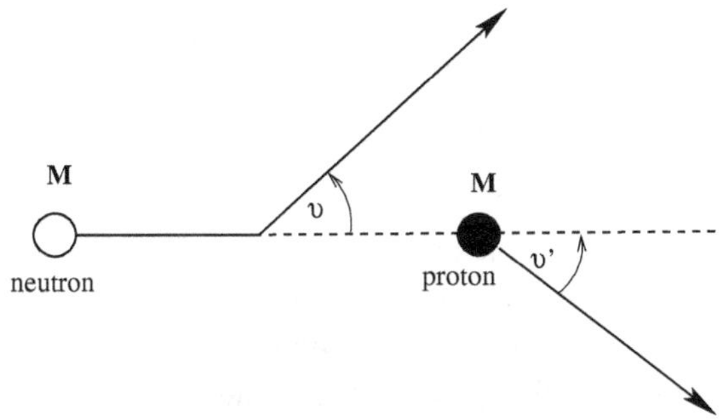

Figure 2.5 Elastic neutron-proton scattering in the laboratory system. Here $\vartheta + \vartheta' = \frac{\pi}{2}$.

Table 2.4 Relations between $\mathcal{L}$ and $\mathcal{MC}$ systems for describing scattering quantities of particles with the same masses.

Name	$\mathcal{L}$	$\mathcal{MC}$
Mass	M	$\frac{1}{2}M$
Kinetic energy	E	$\frac{1}{2}E$
Neutron scattering angle	ϑ	2ϑ
Proton scattering angle	ϑ'	$2\vartheta'$

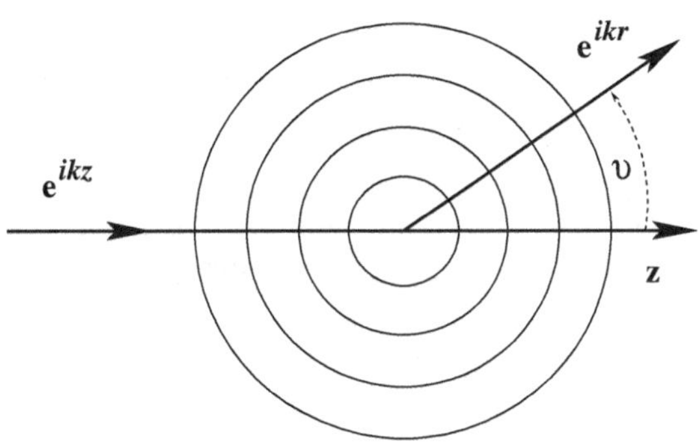

Figure 2.6 Nucleon scattering on central potential.

When $r \to \infty$, the diffuse wave is a superposition of the initial plane-wave e^{ikz} (neutron runs along the z-axis) and the diffused spherical wave e^{ikr}/r , modified depending on the scattering angle ϑ. So, one has

$$\psi_{r\to\infty} = \mathcal{A}\left(e^{ikz} - \frac{e^{ikr}}{r} f(\vartheta)\right). \tag{2.109}$$

ϑ is the angle between the incident particle's trajectory and the scattered particle's direction. In the case of non-central potentials the diffuse wave depends too on the mutual setting of spins of colliding nucleons.

Specifying probability densities $\rho = \psi^\star\psi$ for the incident plane wave $\rho_\mathrm{p} = |\mathcal{A}|^2$ and distributed radial one $\rho_\mathrm{r} = |\mathcal{A}|^2 f(\vartheta)/r^2$, one can evaluate the differential cross section for scattering towards ϑ into a solid angle Ω. The number of scattered particles passing through the dS surface is proportional to

$$d\sigma = \frac{\rho_\mathrm{r}}{\rho_\mathrm{p}} dS = |f(\vartheta)|^2 d\Omega\,, \tag{2.110}$$

where $d\Omega = dS/r^2$.

Hence differential cross-section depends only on the square scattering amplitude $f(\vartheta)$

$$\frac{d\sigma}{d\Omega} = |f(\vartheta)|^2\,, \tag{2.111}$$

and total cross-section will be the full solid angle integral

$$\sigma = \int |f(\vartheta)|^2 d\Omega\,. \tag{2.112}$$

Separating the ψ wave function, which is the solution of the Schroedinger equation with the specified average potential, one will get the radial and angular part of $f(\vartheta)$, and thus find cross sections for the neutron-proton scattering.

If the potential has spherical symmetry, one can apply the partial wave method corresponding to different scattering orbital states l. When the energy of incident particles is low compared to the potential, expansion into partial waves can be limited to small l because it converges quickly.

For model considerations, it is enough to consider the spherical potential in the form of a rectangular well because, in general, asymptotic solutions are not very sensitive to the choice of potential shape. One can use the effective range method if one knows only the average properties of the potential (with a finite range, small in comparison to the de Broglie wavelength of the nucleon).

2.2.1 Partial waves method

Let us consider the Hamiltonian eigenproblem with spherical potential $V(r)$

$$[\hat{T} + V(r)]\psi(r,\vartheta,\varphi) = E\psi(r,\vartheta,\varphi)\,. \tag{2.113}$$

For $E > 0$ this equation can be rewritten in the form

$$(\hat{\Delta} + K^2 + k^2)\psi(r,\vartheta,\varphi) = 0\,, \tag{2.114}$$

where $K^2 \equiv -\frac{MV(r)}{\hbar^2}$, and $k^2 \equiv \frac{ME}{\hbar^2}$. Of course, $k^2 > 0$ for unbound states.

One can separate the functions ψ into radial and angular parts and expand them in a series of so-called partial waves corresponding to the different l values. It is especially favorable for small energies because the contribution of states with large l is small, and the series can be limited to the first few terms.

Let us first examine the expansion of the plane wave e^{ikz} first, being the solution of free particle Hamiltonian eigenproblem,

$$-\frac{\hbar^2}{M}\hat{\Delta}\psi = E\psi\,, \tag{2.115}$$

when the potential $V(r) = 0$. This is a typical wave equation

$$(\hat{\Delta} + k^2)\psi = 0\,, \tag{2.116}$$

whose solution may be written in the form of a function separated into radial and angular parts

$$\psi(r,\vartheta,\varphi) = \frac{U_l(r)}{r}\,Y_{lm}(\vartheta,\varphi)\,. \tag{2.117}$$

One obtains the following equation for the radial function $U_l(r)$

$$\frac{d^2U_l}{dr^2} + \left(k^2 - \frac{l(l+1)}{r^2}\right)U_l = 0\,, \tag{2.118}$$

which has a solution in the form of spherical Bessel functions

$$\frac{U_l(r)}{r} = j_l(kr) \overset{kr \gg l}{\longrightarrow} \frac{\sin\left(kr - \frac{l\pi}{2}\right)}{kr}\,. \tag{2.119}$$

For problems with axial symmetry in Eq. (2.117), only functions $Y_{l0}(\vartheta,\varphi)$ proportional to Legendre polynomials $P_l(\cos\vartheta)$ appear and individual partial waves

$$\psi_l(r,\vartheta) = j_l(kr)P_l(\cos\vartheta)\,, \tag{2.120}$$

and they form a basis in which one develops the incident plane wave

$$e^{ikz} = \sum_l C_l\, j_l(kr) P_l(\cos\vartheta)\,. \tag{2.121}$$

The following procedure allows us to find the expansion coefficients C_l. Substituting $\cos\vartheta \equiv \xi$ one has

$$e^{ikr\xi} = \sum_l C_l\, j_l(kr)\, P_l(\xi)\,. \tag{2.122}$$

Then, one multiplies the last equation by $P_{l'}$ and integrates with respect to ξ:

$$\int_{-1}^{1} P_{l'}(\xi) e^{ikr\xi} d\xi = \sum_l C_l\, j_l(kr) \int_{-1}^{1} P_{l'}(\xi) P_l(\xi)\, d\xi\,. \tag{2.123}$$

Using the condition of orthogonality of Legendre polynomials (2.77) and integrating the left side, one gets

$$P_{l'}(\xi)\frac{e^{ikr\xi}}{ikr}\Big|_{-1}^{1} - \int_{-1}^{1} \frac{e^{ikr\xi}}{ikr}\frac{dP_{l'}(\xi)}{d\xi} d\xi = C_{l'}\, j_{l'}(kr)\frac{2}{2l'+1}\,. \tag{2.124}$$

Afterwards, one integrates the left side of the above equation again, omitting the prime sign

$$P_l(\xi)\frac{e^{ikr\xi}}{ikr}\Big|_{-1}^{1} - \frac{dP_l(\xi)}{d\xi}\frac{e^{ikr\xi}}{(ikr)^2}\Big|_{-1}^{1} + \int_{-1}^{1} \frac{e^{ikr\xi}}{(ikr)^2}\frac{d^2P_l(\xi)}{d\xi^2} d\xi = C_l\, j_l(kr)\frac{2}{2l+1}\,. \tag{2.125}$$

The higher order terms were neglected here. For $r \to \infty$, one only needs to keep the first term on the left side of the last equation

$$P_l(\xi)\frac{e^{ikr\xi}}{ikr}\Big|_{-1}^{1} = \frac{2}{2l+1} C_l \frac{\sin\left(kr - \frac{l\pi}{2}\right)}{kr}\,. \tag{2.126}$$

Inserting Legendre polynomial values at both integration points $P_l(1) = 1$ and $P_l(-1) = (-1)^l = e^{i\pi l}$, one can transform the left side of the equation as follows:

$$\frac{1}{ikr}\left(e^{ikr} - e^{-ikr}e^{i\pi l}\right) = \frac{e^{\frac{i\pi l}{2}}}{ikr}\left(e^{i\left(kr-\frac{l\pi}{2}\right)} - e^{-i\left(kr-\frac{l\pi}{2}\right)}\right) = \tag{2.127}$$

$$\frac{\left(\cos\frac{\pi}{2} + i\sin\frac{\pi}{2}\right)^l}{ikr} 2i\,\sin\left(kr - \frac{l\pi}{2}\right) = 2i^l \sin\left(kr - \frac{l\pi}{2}\right)/kr\,.$$

Comparing this result to the right-hand side of Eq. (2.126)

$$\frac{2\mathrm{i}^l \sin\left(kr - \frac{l\pi}{2}\right)}{kr} = \frac{2}{2l+1} C_l \frac{\sin\left(kr - \frac{l\pi}{2}\right)}{kr}, \tag{2.128}$$

one obtains $C_l = (2l+1)\mathrm{i}^l$. So, for large values kr one can write the plane wave (2.121) in the form

$$\mathrm{e}^{\mathrm{i}kz} = \sum_{l=0}^{\infty} (2l+1)\mathrm{i}^l j_l(kr) P_l(\cos\vartheta). \tag{2.129}$$

Similarly, one can determine the wave function of the eigenproblem of Hamiltonian for particles in the potential well, assuming other partial waves, as radial functions are dependent on the shape of potential $U_l(r)/r$.
Then

$$\psi(r,\vartheta,\varphi) = \sum_{l=0}^{\infty} (2l+1)\mathrm{i}^l \frac{U_l(r)}{r} P_l(\cos\vartheta). \tag{2.130}$$

Because one is interested in the asymptotic form of the ψ function outside the potential well range, one makes simplifying assumptions and adopts U_l in a form analogous to the function j_l modified with so-called phase-shift δ_l

$$\begin{aligned} U_l(r) &\overset{r\to\infty}{\longrightarrow} C_l' \sin\left(kr - \tfrac{l\pi}{2} + \delta_l\right) \\ &= \tfrac{C_l'}{2\mathrm{i}} \left[\mathrm{e}^{\mathrm{i}\left(kr - \frac{l\pi}{2} + \delta_l\right)} - \mathrm{e}^{-\mathrm{i}\left(kr - \frac{l\pi}{2} + \delta_l\right)}\right] \\ &= \mathrm{i}\tfrac{C_l'}{2} \left[\mathrm{e}^{-\mathrm{i}\left(\delta_l - \frac{l\pi}{2}\right)} \mathrm{e}^{-\mathrm{i}kr} - \mathrm{e}^{\mathrm{i}\left(\delta_l - \frac{l\pi}{2}\right)} \mathrm{e}^{\mathrm{i}kr}\right]. \end{aligned} \tag{2.131}$$

Also the expansion (2.129) of the plane wave $\mathrm{e}^{\mathrm{i}kz}$ may be written analogously to (2.131):

$$\begin{aligned} \mathrm{e}^{\mathrm{i}kz} &\overset{r\to\infty}{\longrightarrow} \sum_{l=0}^{\infty} (2l+1)\mathrm{i}^l \frac{\sin\left(kr - \frac{l\pi}{2}\right)}{kr} P_l(\cos\vartheta) \\ &= \sum_{l=0}^{\infty} (2l+1)\frac{\mathrm{i}^{l+1}}{2} \frac{1}{kr} \left(\mathrm{e}^{\mathrm{i}\frac{l\pi}{2}} \mathrm{e}^{-\mathrm{i}kr} - \mathrm{e}^{-\mathrm{i}\frac{l\pi}{2}} \mathrm{e}^{\mathrm{i}kr}\right) P_l(\cos\vartheta). \end{aligned} \tag{2.132}$$

Let us try to express the scattering amplitude $f(\vartheta)$ using the phase shift δ_l. To do it, one inserts into Eq. (2.109) the asymptotic function ψ (2.130) and the plane wave (2.132) expanded into partial waves

$$\begin{aligned} &\sum_{l=0}^{\infty} (2l+1)\mathrm{i}^l \frac{C_l'}{r} \frac{\mathrm{i}}{2} \left[\mathrm{e}^{-\mathrm{i}\left(\delta_l - \frac{l\pi}{2}\right)} \mathrm{e}^{-\mathrm{i}kr} - \mathrm{e}^{\mathrm{i}\left(\delta_l - \frac{l\pi}{2}\right)} \mathrm{e}^{\mathrm{i}kr}\right] P_l(\cos\vartheta) \\ &= \mathcal{A} \sum_{l=0}^{\infty} (2l+1)\frac{\mathrm{i}^{(l+1)}}{2kr} \left[\mathrm{e}^{\mathrm{i}\frac{l\pi}{2}} \mathrm{e}^{-\mathrm{i}kr} - \mathrm{e}^{-\mathrm{i}\frac{l\pi}{2}} \mathrm{e}^{\mathrm{i}kr}\right] P_l(\cos\vartheta) - \mathcal{A} f(\vartheta)\frac{\mathrm{e}^{\mathrm{i}kr}}{r}. \end{aligned} \tag{2.133}$$

This equation is to be fulfilled for every r. So, one must equate to zeros the coefficients standing at e^{-ikr} and at e^{ikr} separately

$$\begin{cases} \sum_{l=0}^{\infty}(2l+1)\frac{P_l(\cos\vartheta)}{r}\frac{i^{l+1}}{2}e^{i\frac{l\pi}{2}}\left[C_l' e^{-i\delta_l}-\frac{\mathcal{A}}{k}\right]=0\,, \\ \sum_{l=0}^{\infty}(2l+1)\frac{P_l(\cos\vartheta)}{r}\frac{i^{l+1}}{2}e^{-i\frac{l\pi}{2}}\left[C_l' e^{i\delta_l}-\frac{\mathcal{A}}{k}\right]-\frac{\mathcal{A}f(\vartheta)}{r}=0\,. \end{cases} \tag{2.134}$$

This set of equations allow to determine the coefficients $C_l' = \mathcal{A}e^{i\delta_l}/k$ and one obtains the formula for the scattering amplitude:

$$f(\vartheta)=\frac{1}{2k}\sum_{l=0}(2l+1)i^{l+1}\left[e^{2i\delta_l}-1\right]P_l(\cos\vartheta)\,, \tag{2.135}$$

which depends on the phase-shift δ_l, i.e. the modification of wave function caused by the potential. For low energy, it is sufficient to consider the S wave only, i.e. the first term with $l = 0$ in the sum (2.135):

$$f(\vartheta)=f_0=\frac{i}{2k}\left(e^{2i\delta_0}-1\right)=-\frac{1}{k}e^{i\delta_0}\sin\delta_0\,. \tag{2.136}$$

Hence the differential cross-section (2.111) is:

$$\frac{d\sigma_0(\vartheta)}{d\Omega}=\frac{1}{k^2}\sin^2\delta_0\,, \tag{2.137}$$

and total one

$$\sigma_0=\frac{4\pi}{k^2}\sin^2\delta_0\,. \tag{2.138}$$

For low energy, i.e., for $k \to 0$ one has $f_0 = a$ and one calls a the scattering length.

The scattering length can be interpreted geometrically as the zero points of the asymptotic form of the function $U_0(r)$. For small energies, $k \to 0$, and out of the potential range, the radial function corresponding to $l = 0$ satisfies equation $rU_0'' = 0$. So,

$$U_0 = r\psi = \mathcal{A}\left(re^{ikz}-f_0\,e^{ikr}\right)\xrightarrow{k\to 0}\mathcal{A}(r-a)\,. \tag{2.139}$$

The dashed line in Fig. 2.7 illustrates this asymptotic form of $U_0(r)$.

Of course, for $k \to 0$, the phase-shift $\delta_0 \to 0$ and then, starting from Eq. (2.136), one can get the relation

$$f_0 = a = \lim_{k\to 0}\left(-\frac{1}{k}e^{i\delta_0}\sin\delta_0\right)=-\frac{\delta_0}{k}\,. \tag{2.140}$$

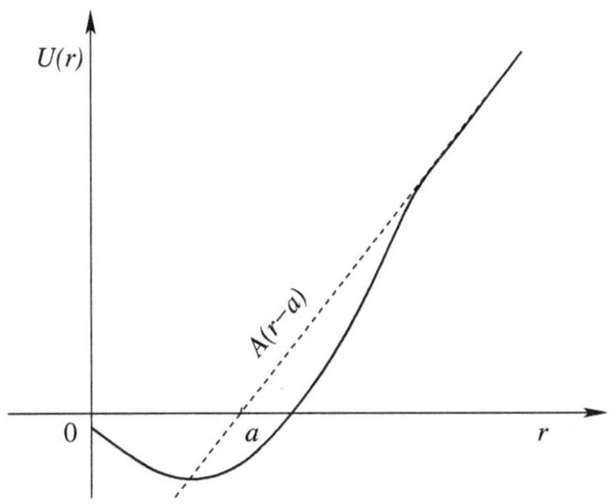

Figure 2.7 Geometric interpretation of the scattering length a.

Hence the phase-shift

$$\delta_0 \overset{k \to 0}{\longrightarrow} -ak\,. \tag{2.141}$$

The size and sign of the scattering length a depend on the potential shape. Using the last expression and Eq. (2.138), one can treat the cross-section for low energy scattering as $\sigma_0 = 4\pi a^2$, i.e., as if the particles were scattered on a sphere with a radius of a.

2.2.2 Neutron-proton scattering on spherical rectangular well potential

One can obtain the exact form of the phase-shift δ_0 and scattering amplitudes needed to determine the total cross-section by solving the Schroedinger equation for a specific average potential, e.g., rectangular well.

One solves the problem in a similar way as in Sec. 2.1.1, remembering, however, that for the scattered states $E > 0$

$$\begin{cases} \left(-\frac{\hbar^2}{M}\hat{\Delta} - V_0\right)\psi_I = E\psi_{\mathrm{I}} & \text{for} \quad r \leq \alpha\,, \\ -\frac{\hbar^2}{M}\hat{\Delta}\psi_{\mathrm{II}} = E\psi_{\mathrm{II}} & \text{for} \quad r > \alpha\,. \end{cases} \tag{2.142}$$

One assumes the potential in a similar form as in Eq. (2.6)

$$V(r) = \begin{cases} -V_0 & \text{for} \quad r \leq \alpha \quad (\mathrm{I})\,, \\ 0 & \text{for} \quad r > \alpha \quad (\mathrm{II})\,. \end{cases} \tag{2.143}$$

This rectangular potential well is shown in Fig. 2.8.

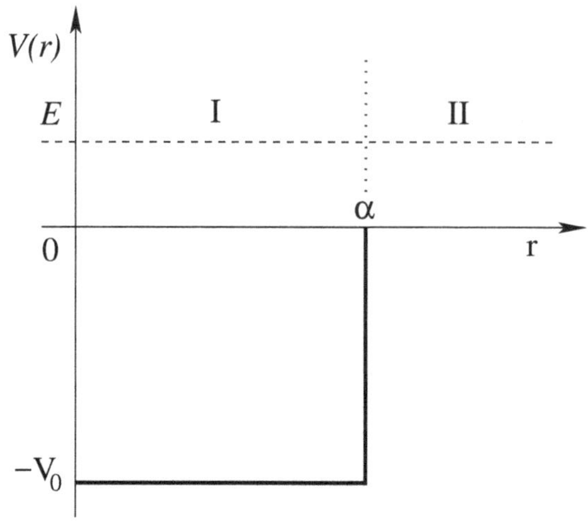

Figure 2.8 Rectangular potential well.

Entering the notation

$$K^2 \equiv \frac{MV_0}{\hbar^2} \quad \text{and} \quad k^2 \equiv \frac{EM}{\hbar^2}\,, \tag{2.144}$$

and

$$\kappa^2 \equiv k^2 + K^2\,, \tag{2.145}$$

one can rewrite the Hamilton equations (2.142) in the form

$$\begin{cases} (\hat{\Delta} + \kappa^2)\psi_{\rm I} = 0 \quad \text{for} \quad r \le \alpha\,, \\ (\hat{\Delta} + k^2)\psi_{\rm II} = 0 \quad \text{for} \quad r > \alpha\,. \end{cases} \tag{2.146}$$

After substitution

$$\psi = \frac{U(r)}{r} Y_{lm}(\vartheta, \phi) \tag{2.147}$$

and taking only the S-wave solution, i.e. $l = 0$, one gets the radial equations in the form

$$\begin{cases} U_{\rm I}'' + \kappa^2 U_{\rm I} = 0\,, \\ U_{\rm II}'' + k^2 U_{\rm II} = 0\,. \end{cases} \tag{2.148}$$

In analogy to (2.119) and (2.131) one can write their asymptotic solutions for large r

$$\begin{cases} U_{\rm I} = \mathcal{A} \sin(\kappa r)\,, \\ U_{\rm II} = \mathcal{B} \sin(kr + \delta_0)\,. \end{cases} \tag{2.149}$$

From the condition of continuity of the logarithmic derivatives (2.31) one has

$$\kappa \cot(\kappa\alpha) = k \cot(k\alpha + \delta_0) . \tag{2.150}$$

From here one can determine the sinus of phase-shift

$$\sin\delta_0 = \frac{\pm \sin(k\alpha)}{\sqrt{k^2 + \kappa^2 \cot^2(\kappa\alpha)}} \left[k\cot(k\alpha) - \kappa\cot(\kappa\alpha) \right] \tag{2.151}$$

which will be used to determine σ_0 from formula (2.138):

$$\sigma_0 = \frac{4\pi}{k^2} \frac{\sin^2(k\alpha)}{\left[k^2 + \kappa^2 \cot^2(\kappa\alpha)\right]} \left[k\cot(k\alpha) - \kappa\cot(\kappa\alpha) \right]^2 . \tag{2.152}$$

For $k \to 0$, i.e. for low energy, $\sin(k\alpha) \approx k\alpha$ and $\kappa \gg k$, so

$$\sigma_0 = 4\pi\alpha^2 . \tag{2.153}$$

Thus, the cross-section is equal to the surface of the sphere with a radius α equal to the potentials' range. E.g., if $\alpha = 4$ fm then $\sigma_0 \approx 2$b, were 1b=1barn=10^{-28} m^2).

2.2.3 Effective range theory

Suppose one is dealing with a central potential with a finite range, small compared to the de Broglie neutron wavelength. In that case, a cross-section can be evaluated using potential parameters estimated from the experimental data.

Let us consider the $V(r)$ potential of α range and two eigenenergies E_i, where $i = 1, 2$, as shown in Fig. 2.9.

By entering notations

$$k_i^2 \equiv \frac{ME_i}{\hbar^2} \quad \text{and} \quad K^2 \equiv \frac{MV(r)}{\hbar^2} , \tag{2.154}$$

for the S-wave ($l = 0$), one gets the following equation for the radial function U_i:

$$U_i'' + (k_i^2 + K^2)\, U_i = 0 . \tag{2.155}$$

Alternatively, considering the case without potential, one evaluates the radial functions $\mathcal{V}_i$ by solving the equation

$$\mathcal{V}_i'' + k_i^2 \mathcal{V}_i = 0 . \tag{2.156}$$

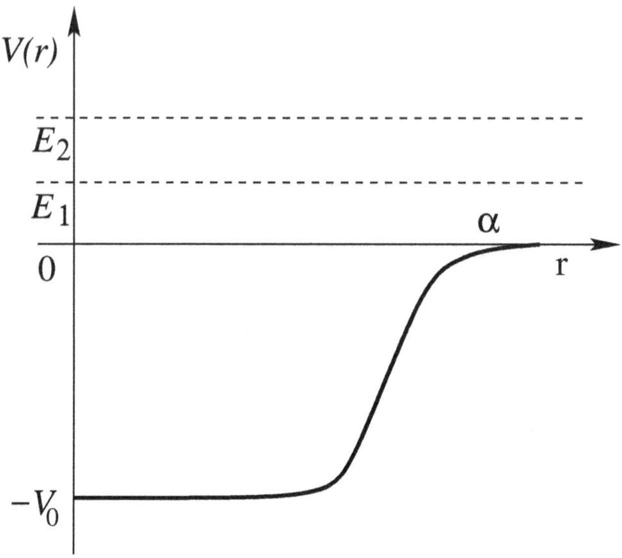

Figure 2.9 Nucleon scattering potential in the effective range theory.

The functions U_i and $\mathcal{V}_i$ have the same asymptotic form

$$\lim_{r\to\infty}[U_i(r) - \mathcal{V}_i(r)] = 0. \tag{2.157}$$

Consider a pair of equations for $i = 1, 2$, multiplied by U_2 and U_1, respectively

$$\begin{cases} U_1'' = -(k_1^2 + K^2)U_1 \quad | \quad \cdot U_2\,, \\ U_2'' = -(k_2^2 + K^2)U_2 \quad | \quad \cdot U_1\,. \end{cases} \tag{2.158}$$

Let us subtract them by sides

$$U_1''\,U_2 - U_2''\,U_1 = (k_2^2 - k_1^2)\,U_1 U_2 \tag{2.159}$$

and integrate both sides

$$\int_0^\infty (U_1'\,U_2 - U_2'\,U_1)'\,dr = (k_2^2 - k_1^2)\int_0^\infty U_1 U_2\,dr. \tag{2.160}$$

Hence

$$(U_1'\,U_2 - U_2'\,U_1)|_0^\infty = (k_2^2 - k_1^2)\int_0^\infty U_1 U_2\,dr. \tag{2.161}$$

Similar transformations can be made for the functions $\mathcal{V}_i$:

$$(\mathcal{V}_1'\mathcal{V}_2 - \mathcal{V}_2'\mathcal{V}_1)|_0^\infty = (k_2^2 - k_1^2)\int_0^\infty \mathcal{V}_1\mathcal{V}_2\,dr. \tag{2.162}$$

Using the boundary condition: $U_i(0) = 0$ and the two above expressions one obtains the following equations:

$$\begin{cases} U_1'(\infty)U_2(\infty) - U_2'(\infty)U_1(\infty) & = \quad (k_2^2 - k_1^2)\int_0^\infty U_1U_2\,dr \\ \mathcal{V}_1'(\infty)\mathcal{V}_2(\infty) - \mathcal{V}_2'(\infty)\mathcal{V}_2(\infty) & = \quad (k_2^2 - k_1^2)\int_0^\infty \mathcal{V}_1\mathcal{V}_2\,dr \\ & + \quad \mathcal{V}_1'(0)\mathcal{V}_2(0) - \mathcal{V}_2'(0)\mathcal{V}_1(0), \end{cases} \tag{2.163}$$

which, after subtacting both sides, gives

$$\mathcal{V}_2'(0) - \mathcal{V}_1'(0) = (k_2^2 - k_1^2)\int_0^\infty (\mathcal{V}_1\mathcal{V}_2 - U_1U_2)\,dr. \tag{2.164}$$

One has assumed here that $U_i(\infty) = \mathcal{V}_i(\infty)$ and replaced normalization of the asymptotic function by the assumption $\mathcal{V}_i(0) = 1$.
The radial equation (2.156) has a solution

$$\mathcal{V}_i = \mathcal{A}e^{ik_ir} + \mathcal{B}e^{-ik_ir}, \tag{2.165}$$

where $\mathcal{A}$ and $\mathcal{B}$ are constants. For further consideration, let us convert the last expression to a more useful form

$$\begin{aligned} \mathcal{V}_i &= \mathcal{A}e^{ik_ir} + \mathcal{B}e^{-ik_ir} = \mathcal{A}'\sin(k_ir) + \mathcal{B}'\cos(k_ir) \\ &= \sqrt{\mathcal{A}'^2 + \mathcal{B}'^2}\left(\frac{\mathcal{A}'}{\sqrt{\mathcal{A}'^2 + \mathcal{B}'^2}}\sin(k_ir) + \frac{\mathcal{B}'}{\sqrt{\mathcal{A}'^2 + \mathcal{B}'^2}}\cos_i(k_ir)\right) = \\ &= \sqrt{\mathcal{A}'^2 + \mathcal{B}'^2}\,\frac{\mathcal{B}'}{\mathcal{B}'}\left[\cos\delta_i\sin(k_ir) + \sin\delta_i\cos(k_ir)\right] \\ &= \mathcal{B}'\,\frac{\sin(k_ir + \delta_i)}{\sin\delta_i}. \end{aligned} \tag{2.166}$$

The condition $\mathcal{V}_i(0) = 1$ gives $B' = 1$, so

$$\mathcal{V}_i(r) = \frac{\sin(k_ir + \delta_i)}{\sin\delta_i} \tag{2.167}$$

and

$$\mathcal{V}_i'(r) = \frac{k_i}{\sin\delta_i}\cos(k_i r + \delta_i), \tag{2.168}$$

i.e.

$$\mathcal{V}_i'(0) = k_i \cot(\delta_i). \tag{2.169}$$

Hence Eq. (2.164) takes the form

$$k_2\cot(\delta_2) - k_1\cot(\delta_1) = (k_2^2 - k_1^2)\int_0^\infty (\mathcal{V}_1\mathcal{V}_2 - U_1U_2)\,dr. \tag{2.170}$$

If E_1 tends to E_2, i.e. when $k_1 \to k_2$, one can rewrite the last equality in the form

$$\frac{d(k\cot\delta)}{d(k^2)} = \int_0^\infty (\mathcal{V}^2 - U^2)\,dr. \tag{2.171}$$

The relationship (2.144) between the length of the wave vector and energy was used here.. Let us now expand the function $k\cot\delta$ into the Taylor series around $E = 0$, i.e. around $k^2 = 0$:

$$\begin{aligned} k\cot\delta &= k\cot\delta|_{k^2=0} + \frac{1}{2}\frac{d(k\cot\delta)}{d(k^2)}|_{k^2=0}k^2 + \ldots \\ &= k\cot\delta|_{k^2=0} + \frac{k^2}{2}\left[\int_0^\infty (\mathcal{V}^2 - U^2)\,dr\right]_{k^2=0}, \end{aligned} \tag{2.172}$$

and define the effective scattering range r_0:

$$r_0 \equiv \left[\int_0^\infty (\mathcal{V}^2 - U^2)\,dr\right]_{k^2=0}. \tag{2.173}$$

One can also recall here the asymptotic solution (2.109) for the distributed function

$$U = A\,r\left(\mathrm{e}^{ikz} - f(\vartheta)\,\mathrm{e}^{ikr}\right) \overset{k\to 0}{=} A(r-a), \tag{2.174}$$

where $f(\vartheta) = a$ was the scattering length. From the condition $\mathcal{V}(0) = 1$ one has $-\mathcal{A}a = 1$, i.e. $\mathcal{A} = -1/a$, hence

$$\mathcal{V}(r) = 1 - \frac{r}{a} \quad \text{for} \quad r \gg \alpha. \tag{2.175}$$

So

$$\mathcal{V}'(0)|_{k^2=0} = -\frac{1}{a} = k\cot\delta|_{k^2=0}. \tag{2.176}$$

The expansion (2.172) for $k\cot\delta$ therefore takes the form

$$k\cot\delta = -\frac{1}{a} + \frac{1}{2} r_0 k^2, \tag{2.177}$$

which allows to express the total cross-section (2.138) in the following form:

$$\sigma = \frac{4\pi}{k^2}\sin^2\delta = \frac{4\pi}{k^2}(1+\cot^2\delta)^{-1} = \frac{4\pi}{k^2 + \left[\frac{1}{2} r_0 k^2 - \frac{1}{a}\right]^2}. \tag{2.178}$$

As can be seen, for medium-energy neutrons, the scattering cross-section is described by two constants associated with global potential properties: scattering length a and effective range r_0. Knowing the experimental cross-sections, one can obtain information on the range and depth of the potential.

To study the dependence of the interaction on the spin of neutrons, the scattering on ortho- and para-hydrogen should be done, i.e., on triplet (a_t, r_{0t}) and singlet (a_s, r_{0s}) states. Since there are three times more triplet states due to three possibilities of spin projection $m_s = 0, \pm 1$, one has

$$\sigma = \frac{3}{4}\sigma_t + \frac{1}{4}\sigma_s, \tag{2.179}$$

where

$$\sigma_t = \frac{4\pi}{k^2 + \left(\frac{1}{2} r_{0t} k^2 - \frac{1}{a_t}\right)^2}, \qquad \sigma_s = \frac{4\pi}{k^2 + \left(\frac{1}{2} r_{0s} k^2 - \frac{1}{a_s}\right)^2}, \tag{2.180}$$

are respectively cross-sections for scattering in the triplet $(S = 1)$ and singlet $(S = 0)$ state.

2.2.4 Proton on proton and neutron-neutron scattering

In comparison to the neutron scattering on protons, which one has already discussed, the proton on proton scattering theory at low energies ($E < 10$ MeV) should include several additional effects:

1. Identity of particles, fermions, which the Pauli principle prohibits occupying more than one quantum state,
2. Coulomb repulsion of protons,
3. No bound states in systems are composed of protons only.

The study of neutron scattering on neutrons is complicated. The neutron target cannot be created, and the reactions available in practice require the use of three bodies to interpret the theory, such as the description of the scattering of neutrons or deuterons on tritons:

$$\begin{aligned} n + {}^3H &\rightarrow n + n + d \\ d + {}^3H &\rightarrow {}^3He + n + n. \end{aligned}$$

In general, corrections which should be added to the presented above methods are related to:

1. Difference in mass of proton and neutron,
2. Smoothing of electric charge on the edge of the nucleus,
3. Electromagnetic interaction of magnetic moments of both nucleons.

2.3 VELOCITY-DEPENDENT FORCES

So far, one has considered nuclear forces and static average potentials, dependent only on position coordinates. This approximation is sufficient for low-energy nucleons. However, when E is greater than 5 MeV it is necessary to add velocity-dependent components. The construction of these components is phenomenological, but the relativistic quantum mechanics can justify it. On the other hand, these components must meet some conditions that limit the class of potentials dependent on velocity:

1. because of the invariance of interaction due to the Galileo transformation, the forces should not depend on the choice of the inertial system, and the interaction can only be dependent on the relative momenta of particles $\hat{\mathbf{p}} = \hat{\mathbf{p}}_1 - \hat{\mathbf{p}}_2$,
2. the interaction must be a scalar, so having the location vector $\mathbf{r}$, momentum $\hat{\mathbf{p}}$, spin $\hat{\sigma}$ and limiting to their first powers, one can build the following scalars:

$$(\mathbf{r} \cdot \hat{\mathbf{p}}),\ (\hat{\sigma} \cdot \hat{\mathbf{p}}),\ (\mathbf{r} \times \hat{\mathbf{p}}) \cdot \hat{\sigma}. \tag{2.181}$$

The scalar $(\mathbf{r} \cdot \hat{\mathbf{p}})$ is not time reversal invariant :

$$\hat{T}(\mathbf{r} \cdot \hat{\mathbf{p}})\hat{T}^{-1} = -(\mathbf{r} \cdot \hat{\mathbf{p}}),$$

because velocity changes sign. Scalar $(\hat{\sigma} \cdot \hat{\mathbf{p}})$ is not invariant relative to the operation of the inverse of spatial coordinates $\hat{P}$:

$$\hat{P}(\hat{\sigma} \cdot \hat{\mathbf{p}})\hat{P}^{-1} = -(\hat{\sigma} \cdot \hat{\mathbf{p}}),$$

because the internal angular momentum $\hat{\sigma}$ does not change the sign due to the operation $\hat{P}$.

All that remains is the coupling of orbital motion with spin in the form:

$$\hat{V}_{\rm LS} = V_{\rm ls}(r)\,\hat{\mathbf{L}}\cdot\hat{\mathbf{S}}, \tag{2.182}$$

where $\hat{\mathbf{S}} = \frac{1}{2}(\hat{\sigma}_1 + \hat{\sigma}_2)$ and $\hat{\mathbf{L}} = (\mathbf{r}\times\hat{\mathbf{p}})$.

Since the nucleon-nucleon interaction is charge-independent and symmetric in isospin space, it must also be symmetric in the spin-spatial part. Symmetry in spin variables occurs in the spin-orbit and the tensor potentials. It means that these potentials commute with the operator $\hat{\mathbf{S}}^2$, so spin S is a good quantum number, and both types of potentials do not couple singlet and triplet states. Therefore, there is an analog element in nuclear interaction like that spin-orbit coupling in atoms created by a circulating electron. Whereas one can easily get such potential for an atom based on the electromagnetic coupling of orbital and spin magnetic momenta of an electron, in the majority of nuclear models it has only a phenomenological character. The spin-orbit interaction can be derived from the more fundamental but also phenomenological, relativistic mean-field theory (RMFT, see Sec. 7.5).
The nuclear spin-orbit potential is usually taken in the following form:

$$\hat{V}_{\rm LS} = \frac{-1}{2M^2c^2}(\mathbf{grad}\,V\times\hat{\mathbf{p}})\cdot\hat{\sigma}. \tag{2.183}$$

For the spherical central potential, one has:

$$\mathbf{grad}\,V = \frac{\mathbf{r}}{r}\frac{\partial V}{\partial r}, \tag{2.184}$$

and

$$\hat{V}_{LS} = \frac{-1}{2M^2c^2}\,\frac{1}{r}\frac{\partial V}{\partial r}\hat{\mathbf{L}}\cdot\hat{\mathbf{S}}. \tag{2.185}$$

Goeppert-Mayer and Jensen in 1950 first introduced the spin-orbit term in nuclear physics to explain the origin of the magic numbers in nuclei. They have added this term to the mean single-particle potentials with parameters adjusted to experimental data. Some authors also consider terms with higher powers $\hat{\mathbf{p}}$ in the nucleon-nucleon potentials.

CHAPTER 3

Nuclear matter

3.1 DEFINITION OF NUCLEAR MATTER

In this chapter one is going to pass from the two-body problem to the nuclear matter theory. The nuclear matter is an invented system with an equal but infinite number of protons and neutrons ($N = Z$), having constant mass and charge densities. The interactions between nucleons in this matter are strong but saturated, ensuring constant binding energy for one nucleon. One makes also an additional simplifying assumption that the Coulomb interaction does not act between the protons. One assumes in addition that the forces operating in such a system consist of the two-body interactions only. One takes also into account the effect of the Pauli exclusion principle in the nuclear matter. In the following, one will concentrate mostly on these aspects of the nuclear matter theory which lead to the justification of the single-particle model.

We assume that interaction in the matter consisting of A nucleons is the sum of two-body forces depending on $\{\mathbf{x}_i\} \equiv$ position $\mathbf{r}_i$, spins $\mathbf{s}_i$ and isospins $\mathbf{t}_i$

$$V(\mathbf{r}_1, \mathbf{r}_2, \ldots, \mathbf{r}_A, \mathbf{s}_1, \mathbf{s}_2, \ldots, \mathbf{s}_A, \mathbf{t}_1, \mathbf{t}_2, \ldots, \mathbf{t}_A) = \sum_{i,j=1}^{A} V_{ij}. \tag{3.1}$$

The wave function of the system (Ψ) is taken as a product of the individual nucleon functions $\varphi_{\nu_i}(\mathbf{x}_i) = \varphi_{\nu_i}(\mathbf{r}_i, \mathbf{s}_i, \mathbf{t}_i)$

$$\Psi = \prod_{i=1}^{A} \varphi_{\nu_i}(\mathbf{x}_i) = \prod_{i=1}^{A} \psi(\mathbf{r}_i)\chi(\mathbf{s}_i)\eta(\mathbf{t}_i) \tag{3.2}$$

consisting of the spatial ψ, spin χ and isospin η parts. The half-number $\frac{1}{2}\hbar$ spin of nucleons, i.e. fermions, decides that they are subject to the Fermi-Dirac statistics and the systems' wave-function should be asymmetric with respect the exchange of particles. One can achieve it, e.g., by taking Slater determinant built from single-particle functions in different states ν_i:

$$\Psi = \frac{1}{\sqrt{A!}} \begin{vmatrix} \varphi_{\nu_1}(\mathbf{x}_1) & \varphi_{\nu_2}(\mathbf{x}_1)\ldots & \varphi_{\nu_A}(\mathbf{x}_1) \\ \varphi_{\nu_1}(\mathbf{x}_2) & \varphi_{\nu_2}(\mathbf{x}_2)\ldots & \varphi_{\nu_A}(\mathbf{x}_2) \\ \cdot & \cdot & \cdot \\ \cdot & \cdot & \cdot \\ \cdot & \cdot & \cdot \\ \varphi_{\nu_1}(\mathbf{x}_A) & \varphi_{\nu_2}(\mathbf{x}_A)\ldots & \varphi_{\nu_A}(\mathbf{x}_A) \end{vmatrix} \tag{3.3}$$

This procedure becomes complicated when the mass-number A increases, and for an infinite system it is impossible to realize. However, by analyzing the solution of the Schroedinger equation in the case of two-body potential acting between two nucleons surrounded by other particles, one can get interesting analogies between wave functions of free nucleons and those in the nuclear matter.

3.2 'HEALING' OF TWO NUCLEONS' WAVE FUNCTION IN NUCLEAR MATTER

Let us consider two nucleons in vacuum having momenta $\mathbf{p}_1, \mathbf{p}_2$ and interaction via the V_{12} potential. Such a system fulfill the following Hamilton equation:

$$(\hat{T}_1 + \hat{T}_2 + V_{12} - E)\psi_S = 0. \tag{3.4}$$

If one puts this pair into the nuclear matter, one has to add a projection operator $\hat{Q}_{12}$, which will eliminate from a dispersed wave the states with moments smaller than the Fermi one. It is because all states except the initial ones $\mathbf{p}_1$ and $\mathbf{p}_2$ are already occupied by other nucleons. Then one gets the Bethe-Goldstone equation:

$$(\hat{T}_1 + \hat{T}_2 + \hat{Q}_{12} V_{12} - E)\psi_{BG} = 0, \tag{3.5}$$

which will be discussed in the next chapter. Consider, for comparison, the equation for non-interacting free particles

$$(\hat{T}_1 + \hat{T}_2 - E)\psi_0 = 0, \tag{3.6}$$

which has a solution in the form of a product of the plane-wave functions.

Without going into the analytical solutions of Eqs. (3.4)-(3.6), let us compare the results presented schematically in Fig. 3.1. The scattering effect appears in a phase shift of the function ψ_S relative to the plane wave function.

However, it turns out that the function ψ_{BG} is similar to ψ_S (scattering in vacuum) over short distances. After exceeding x_{hd} which is called healing distance, it becomes almost identical with the plane-wave function ψ_0, obtained without interaction. It means that nuclear forces modify the function only over short distances. Whereas, beyond the healing distance, the function runs as if there was no interaction.

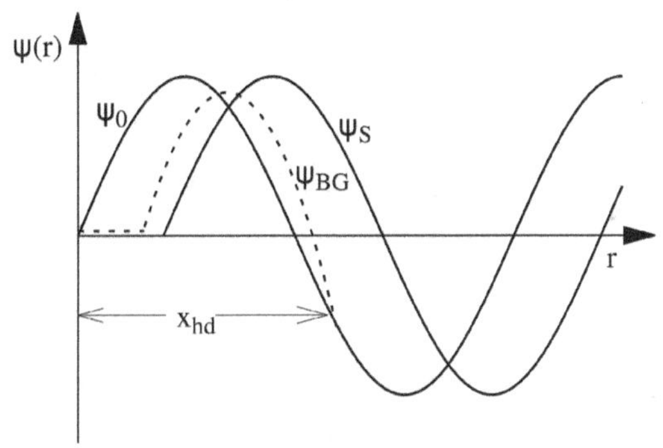

Figure 3.1 Wave functions of a pair of nucleons: interacting in a vacuum ψ_S, in nuclear matter ψ_{BG}, and non-interacting ψ_0, in the function of the relative distance $\mathbf{r} = \mathbf{r}_1 - \mathbf{r}_2$ between particles.

The experimental data show that the average free path of nucleon $\bar{l}$ in the nucleus is greater than the healing distance $\bar{l} > x_{hd}$. Namely, the average free path of a nucleon in a nucleus can be estimated by analyzing the experimental total cross sections for scattering. Fig. 3.2 shows the energy dependence of total cross sections of neutrons scattered on lead, cadmium, and copper, assuming that the average freeway is close to the radius of the nucleus $\bar{l} \approx R = 1.4A^{1/3}$ fm (solid line) or that $\bar{l} \ll R$ a lot smaller than nucleus size (dashed line). On the average the free path of nucleon is small compared to the radius nuclei, and the total active cross-section changes monotonically with the energy increase up to the value of $2\pi R^2$. The fluctuations of the solid curves are due to interference between the ingoing and outgoing waves for a long free path. Experimental points, not marked here, would simply lie on these curves. It means that the nucleons never get close together on such a distance in which their wave function is *sensitive* on interaction, i.e., they behave like free particles in an average potential well. For heavy nuclei, one can ignore surface effects and get a model of free nucleons – Fermi gas.

To prove the independent particle model's applicability, one can also make simple considerations about the conservation of momentum and energy of two nucleons colliding in the nucleus, neglecting the correlations and the higher-order perturbation terms and only using the Pauli principle.

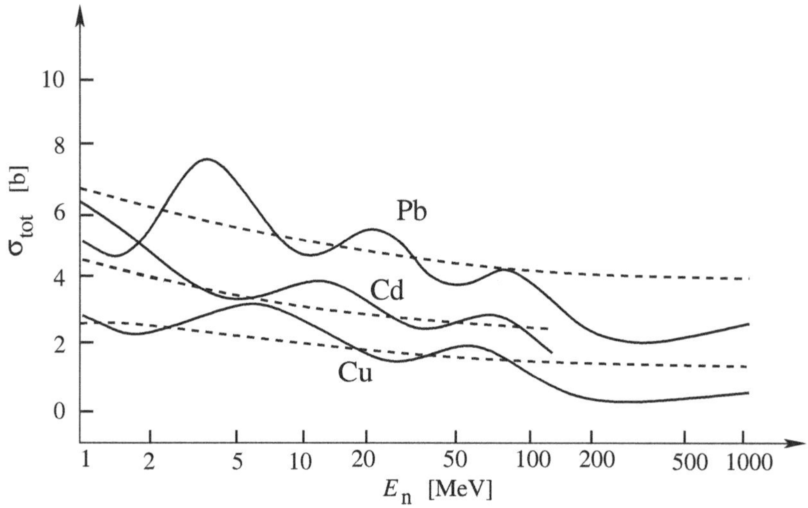

Figure 3.2 Total cross-sections (σ_{tot}) for scattering neutrons on lead (Pu), cadmium (Cd), and copper (Cu) as function neutron energy (E_{n}) for the free-path $\bar{l} \ll R$ (dashed lines) or $\bar{l} \approx R$ (solid lines). Here R is the nucleus radius.

In Fig. 3.3 an example of a single-particle energy spectrum and the Fermi level position e_{F} are shown. All levels with energy lower than e_{F} are occupied. Let us assume two nucleons in the states m and n collide with each other, as a result of which they can stay in their places, exchange places, or occupy free states a and b above e_F because other particles already occupy the neighboring states. The principles of conservation of momentum and energy show that $\mathbf{p}_i$ and energies e_i must meet the relationship

$$\begin{cases} \mathbf{p}_a + \mathbf{p}_b = \mathbf{p}_m + \mathbf{p}_n\,, \\ e_a + e_b = e_m + e_n \end{cases} \tag{3.7}$$

and at the same time there must be fulfilled the inequality $e_m + e_n < e_a + e_b$. As one can see, this is impossible. So, the collision of one type particles does not change their states, i.e. they behave as if the interaction between them does not exist.

3.3 BETHE-GOLDSTONE EQUATION

We have shown in the previous section that the Pauli's exclusion principle switches off effectively the strong interaction between nucleons in nuclei.

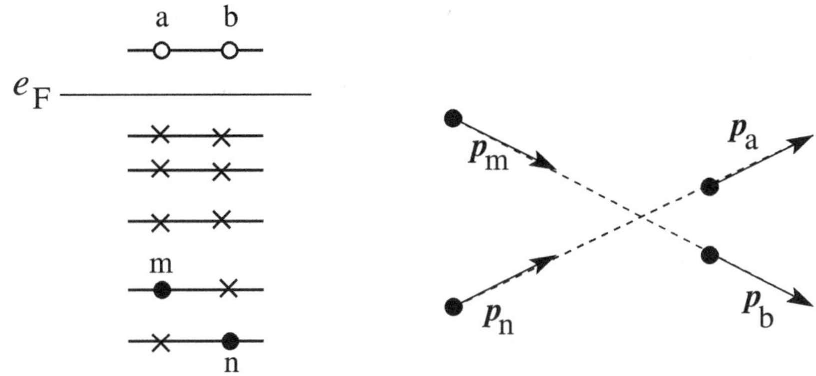

Figure 3.3 Collision of nucleons inside of a nucleus.

Hence, one can roughly describe particles with independent wave functions $\varphi_\nu(\mathbf{x}_i) \equiv \varphi_\nu(i)$, which are the solutions of the single-particle Hamilton eigenproblem

$$\hat{H}_0(i)\varphi_\nu(i) = [\hat{T}(i) + V(i)]\,\varphi_\nu(i) = e_\nu\,\varphi_\nu(i)\,. \tag{3.8}$$

Let's derive the Bethe-Goldstone equation (3.5) for two nucleons immersed in nuclear matter. Disregarding three- and multi-body correlations in the independent pairs approximation, one gets the following Schroedinger equation for two nucleons

$$[\hat{H}_0(1) + \hat{H}_0(2) + V_{12}]\psi_{kl}(1,2) = e_{kl}\,\psi_{kl}(1,2)\,. \tag{3.9}$$

Due to the two-body interaction V_{12}, energy e_{kl} is not a sum $e_k + e_l$. The function ψ_{kl} can be expanded in the base φ_ν in such a way that it will not contain the states occupied by other particles: $m, n, \ldots$ but only empty single-particle states: $a, b, c, \ldots$

$$\begin{aligned} \psi_{kl}(1,2) \;&=\; \varphi_k(1)\varphi_l(2) + \sum_{a,b,a\neq b} C_{ab}\,\varphi_a(1)\,\varphi_b(2) \\ &+\; \sum_a C_{ka}\,\varphi_k(1)\,\varphi_a(2) + \sum_a C_{al}\,\varphi_a(1)\,\varphi_l(2)\,. \end{aligned} \tag{3.10}$$

After inserting the expansion of function ψ_{kl} (3.10) into Eq. (3.9) one gets

$$\begin{aligned} (e_k + e_l)\,\varphi_k(1)\,\varphi_l(2) \;&+\; \sum_{\substack{\alpha,\beta,\alpha\neq\beta;\\ \alpha,\beta\neq k,l}} C_{\alpha\beta}(e_\alpha + e_\beta)\,\varphi_\alpha(1)\,\varphi_\beta(2) \\ &+\; V_{12}\,\psi_{kl}(1,2) = e_{kl}\,\psi_{kl}(1,2)\,. \end{aligned} \tag{3.11}$$

Here, the summation takes place over the states k, l and the empty states: $\alpha = k, a, b, \ldots$ and $\beta = l, a, b, c, \ldots$ (see Fig. 3.3).

Multiplying Eq. (3.11) on the left side by $\varphi_\alpha(1)\varphi_\beta(2)$ and integrating over $d\mathbf{x}_1 d\mathbf{x}_2$, one gets the equation, which in the Dirac notation has the form

$$C_{\alpha\beta}(e_{kl}-e_\alpha-e_\beta) = \langle\varphi_\alpha(1)\varphi_\beta(2)|V_{12}|\psi_{kl}(1,2)\rangle \equiv \langle\alpha,\beta|V|\psi_{kl}\rangle. \qquad (3.12)$$

From here one can already get the expansion coefficients

$$C_{\alpha\beta} = \frac{\langle\alpha\beta|V|\psi_{kl}\rangle}{e_{kl}-e_\alpha-e_\beta} \qquad (3.13)$$

and the wave function (3.10) in the following form:

$$|\psi_{kl}(1,2)\rangle = |kl\rangle + \sum_{\substack{\alpha,\beta,\alpha\neq\beta;\\ \alpha,\beta\neq k,l}} \frac{\langle\alpha\beta|V|\psi_{kl}\rangle}{(e_{kl}-e_\alpha-e_\beta)}|\alpha\beta\rangle. \qquad (3.14)$$

Introducing a single-particle Hamiltonian for two nucleons

$$\hat{H}_0 \equiv \hat{H}_0(1)+\hat{H}_0(2)\,, \quad \text{where} \quad H_0|\alpha\beta\rangle = (e_\alpha+e_\beta)|\alpha\beta\rangle, \qquad (3.15)$$

one can get the function (3.14) in the form

$$|\psi_{kl}(1,2)\rangle = |kl\rangle + \sum_{\substack{\alpha,\beta,\alpha\neq\beta;\\ \alpha,\beta\neq k,l}} \frac{\langle\alpha\beta|V|\psi_{kl}\rangle}{e_{kl}-\hat{H}_0}|\alpha\beta\rangle. \qquad (3.16)$$

Let us define projection operator $\hat{Q}_{kl}$ (3.5)

$$\hat{Q}_{kl} = \sum_{\substack{\alpha,\beta,\alpha\neq\beta;\\ \alpha,\beta\neq k,l}} |\alpha\beta\rangle\langle\alpha\beta| = \sum_{ab}|ab\rangle\langle ab| + \sum_a |ka\rangle\langle ka| + \sum_b |bl\rangle\langle bl| \qquad (3.17)$$

responsible for implementing the Pauli principle in the system. Then using Eq. (3.16) one obtains the Bethe-Goldstone equation in the following form:

$$|\psi_{kl}\rangle = |kl\rangle + \frac{\hat{Q}_{kl}}{e_{kl}-\hat{H}_0}V|\psi_{kl}\rangle. \qquad (3.18)$$

In the nuclear matter in the independent pairs approximation and thanks to the momentum conservation $\mathbf{p} = \mathbf{p}_a + \mathbf{p}_b$, the projection operator (3.17) reduces to the first term only:

$$\hat{Q}_{kl} = \sum_{ab}|ab\rangle\langle ab|, \qquad (3.19)$$

because

$$\mathbf{p}_k + \mathbf{p}_a \neq \mathbf{p} \ \text{ and } \ \mathbf{p}_l + \mathbf{p}_b \neq \mathbf{p}. \qquad (3.20)$$

Let us define the effective $\hat{G}$ interaction which acts on the product of independent functions and whose effect is equivalent to the action of V on the function ψ_{kl} of the system

$$\hat{G}\,|kl\rangle = V|\psi_{kl}\rangle\,. \tag{3.21}$$

Multiplying the Bethe-Goldstone equation (3.18) by V and substituting the definition (3.21), one gets

$$\hat{G}\,|kl\rangle = V|kl\rangle + V\frac{\hat{Q}_{kl}}{e_{kl}-\hat{H}_0}\hat{G}\,|kl\rangle\,, \tag{3.22}$$

hence the interaction must effectively satisfy the equation

$$\hat{G} = V + V\frac{\hat{Q}_{kl}}{e_{kl}-\hat{H}_0}\hat{G}\,. \tag{3.23}$$

Let us now extend our considerations to a many-body system and consider a system of A particles, which in the independent particle model is described by a Hamiltonian

$$\hat{H} = \sum_{i=1}^{A} T(i) + \frac{1}{2}\sum_{i,j=1}^{A} V_{ij} \tag{3.24}$$

and eigenequation

$$\hat{H}\Psi = E\Psi\,. \tag{3.25}$$

Parallel, one can consider the independent particle model with the Hamiltonian

$$\hat{H}_0 = \sum_{i=1}^{A} \hat{H}_0(i)\,, \tag{3.26}$$

having the eigenequation in the form

$$\hat{H}_0\,\Phi_j = W_j\Phi_j\,, \tag{3.27}$$

where the lowest energy corresponding to Φ_0 is

$$W_0 = \sum_{k=1}^{A} e_k\,. \tag{3.28}$$

The eigenfunction Ψ being the solution of Eq. (3.25) can be expanded into wave functions of the independent particles Hamiltonian

$$\Psi = \Phi_0 + \sum_{j=1}^{A} C_j\Phi_j\,. \tag{3.29}$$

Let us derive now the Bethe-Goldstone equation for the system of A nucleons. Let the residual interaction $V_{\rm res}$ be part of the sum of two-body forces, not included in the average potential

$$V_{\rm res} = \frac{1}{2}\sum_{i,j=1}^{A} V_{ij} - \sum_i V(i). \tag{3.30}$$

Then the Hamiltonian (3.24) will be the sum of the independent particles Hamiltonian and the residual interaction

$$\hat{H} = \hat{H}_0 + V_{\rm res} \tag{3.31}$$

and eigenequation (3.25) will take the form

$$(\hat{H}_0 + V_{\rm res})(\Phi_0 + \sum_i C_i \Phi_i) = E(\Phi_0 + \sum_i C_i \Phi_i). \tag{3.32}$$

Multiplying Eq. (3.32) on the left by Φ_0 and integrating over $d\mathbf{x}_1, \ldots, d\mathbf{x}_A$, one gets

$$E = W_0 + \langle \Phi_0 | V_{\rm res} | \Psi \rangle. \tag{3.33}$$

Introducing a multi-body projection operator

$$\hat{Q} \equiv 1 - |\Phi_0\rangle\langle\Phi_0| \tag{3.34}$$

and an effective interaction of the system of A nucleons

$$\hat{G} = V_{\rm res} - V_{\rm res}\frac{\hat{Q}}{E - \hat{H}_0}\hat{G}, \tag{3.35}$$

one obtains the Bethe-Goldstone equation in the following form:

$$\Psi = \Phi_0 + \frac{\hat{Q}}{E - \hat{H}_0} V_{\rm res}\Psi. \tag{3.36}$$

This equation is the exact formulation of the many-body problem, and its solution is one of the main challenges of nuclear and nuclear matter theory. There are several approximate methods to solve it, e.g., the model of independent pairs for two-body short- or/and long-range forces. One uses such methods to describe finite nuclei and neutron stars and one obtains many exciting results for various effective interactions. However, one has to mention here that these results give only corrections to the single-particle approach. The last approach can also be simplified, e.g., by attaching effective two-body forces to the mean-field potential what will be presented in the next chapters.

CHAPTER 4

Independent particles model

4.1 NUCLEAR SIZES

The shape of the average potentials is related to the size and deformation of a nucleus. Due to the short range of nuclear forces, the density distributions of protons and neutrons are generally close to each other. The charge (proton) distribution may be determined experimentally by scattering high-energy electrons on nuclei or, e.g., by examining the hyperfine structure of electron levels in an atom and analyzing α decays.

An example of the scattering of electrons on a gold ($Z = 79$) foil is shown in Fig. 4.1, where the experimental differential cross-section $d\sigma/d\Omega$ (points) depending on scattering angle Θ is compared to its theoretical estimates obtained for assumed different charge distributions. As one can see, in the case of point distribution, theoretical results (dotted line) do not agree with the experimental data (dots). They are better reproduced by assuming a uniform charge distribution (dashed line). One obtains an almost perfect estimate when assuming the charge density distribution in the Fermi function (solid line). The last two distributions are presented in Fig. 4.2.

The uniform charge distribution ρ_u is described as follows:

$$\rho_u = \begin{cases} \rho_0 & \text{for } r \le R, \\ 0 & \text{for } r > R. \end{cases} \tag{4.1}$$

The Fermi distribution ρ_F is similar to the uniform one, but it is smoothed at the edge

$$\rho_F = \rho_0 \left[1 + e^{\frac{r-R}{a}} \right]^{-1}. \tag{4.2}$$

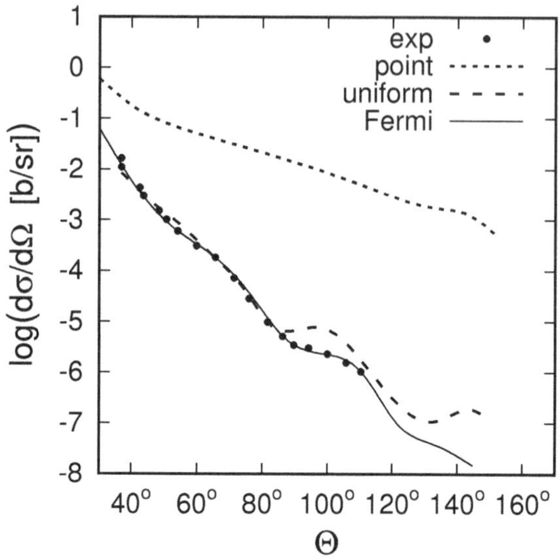

Figure 4.1 Differential cross section for electron scattering on gold obtained theoretically for different charge distributions: point charge (dotted line), uniform (dashed line), and in the form of Fermi functions (solid line) in comparison with experimental data (points).

Fitting theoretical curves to experimental points obtained for many isotopes allows to determine the parameters of nuclear density

$$\rho_0 = 0.17 \frac{\text{nucleon}}{\text{fm}^3} \quad \text{and} \quad a = 0.54\,\text{fm}. \tag{4.3}$$

The a surface smoothing parameter specifies the thickness of the layer surface of nucleus t (see Fig. 4.2), in which the density charge falls from 90% to 10% of ρ_0 ($t \approx 4.4\ a$). The distribution of nucleon density in the nucleus must satisfy the normalization condition

$$A = \int \rho_F d^3 r = \rho_0 \frac{4\pi}{3} R^3 \left[1 + \pi^2 \left(\frac{a}{R} \right)^2 \right], \tag{4.4}$$

Hence, a radius can be obtained for the given ρ_0 and a density distribution (4.2) in the form:

$$R = 1.12 A^{1/3} - 0.86 A^{-1/3} \ldots \text{fm}. \tag{4.5}$$

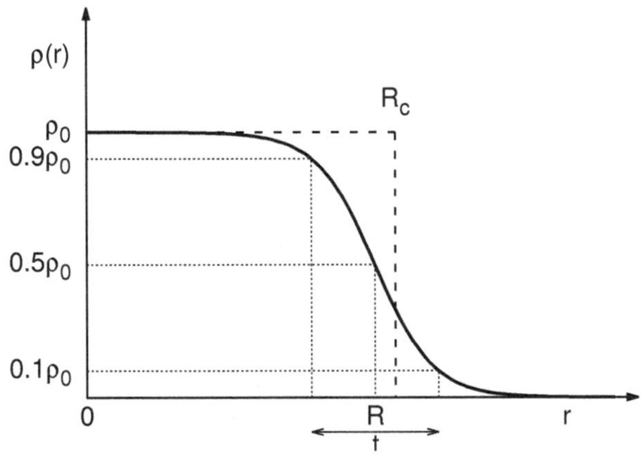

Figure 4.2 Uniform distribution of the charge in the nucleus (dashed line) and Fermi distribution (solid line).

The second term slightly disturbs the commonly assumed proportionality of $R \sim A^{1/3}$. Performing similar integrals, one can evaluate radial moments of density

$$\langle r^n \rangle = \frac{\int \rho_F(r) r^n d^3 r}{\int \rho_F(r) d^3 r} \approx \frac{3}{n+3} R^n \left[1 + \frac{n(n+5)}{6} \pi^2 \left(\frac{a}{R} \right)^2 + \ldots \right]. \tag{4.6}$$

The mean-square radius is

$$\langle r^2 \rangle = \frac{3}{5} (1.12 A^{1/3})^2 \left[1 + 3.84 A^{-2/3} + \ldots \right] \mathrm{fm}^2 \approx \frac{3}{5} (1.2 A^{1/3})^2 \mathrm{fm}^2. \tag{4.7}$$

The effective Coulomb radius R_C can also be estimated from the electrostatic energy of the nucleus. It is the radius of a uniformly charged sphere having the same Coulomb energy as the calculated energy for smoothed charge distribution (4.2):

$$\frac{1}{R_C} = \frac{5}{6} \frac{\int \int \frac{\rho_F(\mathbf{r}_1) \rho_F(\mathbf{r}_2)}{|\mathbf{r}_1 - \mathbf{r}_2|} d^3 r_1 d^3 r_2}{\left[\int \rho_F(r) d^3 r \right]^2} = \frac{1}{R} \left[1 - \frac{7}{6} \pi^2 \left(\frac{a}{R} \right)^2 + \ldots \right], \tag{4.8}$$

which gives approximately

$$R_C \approx 1.2 A^{1/3}\ \mathrm{fm} \quad \text{for} \quad A \approx 100. \tag{4.9}$$

However, this is not a universal relationship but an estimate of the nucleus radius equivalent to a homogeneous charge distribution.

In addition to proton density, one is often interested in mass distribution in the nucleus. In the case of heavy nuclei, where strong Coulomb repulsion and neutron surpluses occur, one needs to consider the distribution's isospin dependence. Some information about it can be obtained from isotope shifts in atomic spectra of π atoms or scattering anti-protons on nuclei.

Nuclear matter's constant density indicates a finite range of nuclear forces, smaller than the nucleus's radius. It leads to the known fact that the binding energy of an atomic nucleus is proportional to its mass number A.

4.2 FERMI GAS MODEL

It is a simple, non-interacting fermions model, solvable analytically, providing a convenient starting point for many estimates regarding, e.g., nuclear binding energy, symmetry energy, distances between single-particle levels, or even shell corrections illustrating the impact of quantum effects on binding energy.

4.2.1 Fermi's energy and the Fermi momentum

Let us assume that the particles move freely in a cubic box L that can be treated as a cube of nuclear matter with a certain number of particles N. There is no position-dependent potential here. The kinetic energy of the particle is, therefore, constant. However, one imposes a periodicity condition on wave functions to transition to a comprehensive nuclear matter system with properties repeated periodically. Fig. 4.3 shows a cube of nuclear matter with the edge L and the volume $L^3 = \mathscr{V}$, in which there are N nucleons.

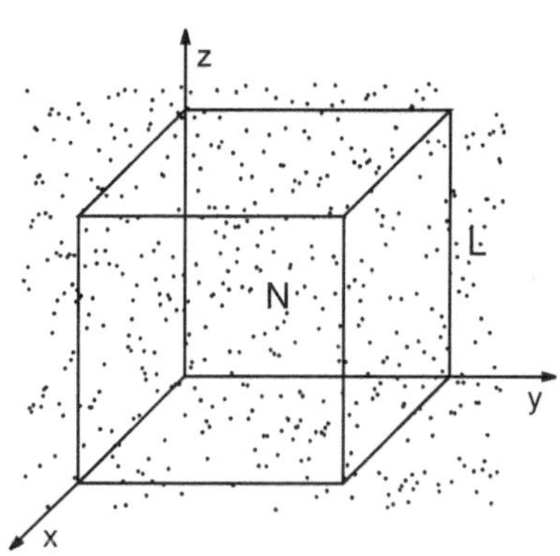

Figure 4.3 Fermi gas model.

Single-particle Schoedinger equation

$$\hat{H}\psi(x,y,z) = e\psi(x,y,z) \tag{4.10}$$

can be separated, assuming that the wave function has product form

$$\psi(x,y,z) = X(x)Y(y)Z(z)\,. \tag{4.11}$$

$\hat{H}$ is simply the kinetic energy operator here

$$-\frac{\hbar^2}{2M}\left(\frac{\partial^2}{\partial x^2} + \frac{\partial^2}{\partial y^2} + \frac{\partial^2}{\partial z^2}\right)\psi = e\psi\,. \tag{4.12}$$

A simple transformation shows that the Schroedinger equation can be separated into the sum of only one type of variables

$$\frac{1}{X}\frac{\partial^2 X}{\partial x^2} + \frac{1}{Y}\frac{\partial^2 Y}{\partial y^2} + \frac{1}{Z}\frac{\partial^2 Z}{\partial z^2} + \frac{2Me}{\hbar^2} = 0\,. \tag{4.13}$$

By comparing individual terms depending on variables x, y, z to constants e_1, e_2, e_3 and denoting $\mathbf{k} = [k_1, k_2, k_3]$

$$k^2 \equiv \frac{2Me}{\hbar^2} \quad \text{and} \quad k_i^2 = \frac{2Me_i}{\hbar^2}\,, \tag{4.14}$$

one obtains for each of the three directions the following type equation:

$$\frac{\partial^2}{\partial x^2}X + k_1^2 X = 0\,, \tag{4.15}$$

whose solution is in the form of $X \sim e^{\alpha x}$, where $\alpha^2 = -k_1^2$, and $\alpha = \pm ik_1$. Let us choose a plane wave running in the direction of x, which gives

$$X(x) = e^{ik_1 x}, \text{ by analogy } Y(y) = e^{ik_2 y} \text{ and } Z(z) = e^{ik_3 z}\,. \tag{4.16}$$

The whole ψ function takes the form

$$\psi(x,y,z) = C\,e^{i(k_1\cdot x + k_2\cdot y + k_3\cdot z)} = C\,e^{i(\mathbf{k}\cdot\mathbf{r})}\,, \tag{4.17}$$

where the wave vector $\mathbf{k}$ has the components k_1, k_2, k_3.
One determines the normalization constant C from the condition of normalizing functions Ψ

$$\iiint_{\mathcal{V}} \psi^*\psi\, d^3r = 1\,, \tag{4.18}$$

or

$$C^2 \int_0^L dx \int_0^L dy \int_0^L \psi^*(x,y,z)\psi(x,y,z)\,dz = 1\,, \tag{4.19}$$

whence

$$C = \frac{1}{\sqrt{\mathscr{V}}}\,. \tag{4.20}$$

Assumed conditions of periodicity in nuclear matter:

$$\psi(x+L,y,z) = \psi(x,y+L,z) = \psi(x,y,z+L) = \psi(x,y,z) \tag{4.21}$$

quantify eigenvalues k_i. For example, for the variable x one has

$$\begin{aligned} &\mathrm{e}^{\mathrm{i}k_1 x} = \mathrm{e}^{\mathrm{i}k_1(x+L)}\,,\\ &\mathrm{e}^{\mathrm{i}k_1 L} = 1\,,\\ &\cos(k_1 L) + \mathrm{i}\sin(k_1 L) = 1\,, \end{aligned} \tag{4.22}$$

hence

$$k_1 L = 2\pi n_1\,, \tag{4.23}$$

where $n_1 = 0, \pm 1, \pm 2, \ldots$. Similar considerations can be carried out in other directions. Therefore:

$$k_i = \frac{2\pi n_i}{L} \quad \text{for} \quad i = 1,2,3\,. \tag{4.24}$$

The length of the vector **k** is therefore quantized:

$$\mathbf{k}^2 = \left(\frac{2\pi}{L}\right)^2 \mathbf{n}^2\,, \tag{4.25}$$

where **n** is a vector with components n_1, n_2, n_3.
The corresponding energy eigenvalue e_n can be obtained using the relation (4.14)

$$e_n = \frac{4\pi^2\hbar^2}{L^2 2M} n^2\,, \tag{4.26}$$

where

$$n^2 = n_1^2 + n_2^2 + n_3^2\,.$$

The eigenfunction has the form

$$\psi_n = \frac{1}{\sqrt{\mathscr{V}}}\mathrm{e}^{\mathrm{i}\mathbf{k}\cdot\mathbf{r}}\,. \tag{4.27}$$

For particles with specified spin **s** and isospin **t** single-particle wave function has the form

$$\varphi_\nu = \frac{1}{\sqrt{\mathscr{V}}}\, \mathrm{e}^{\mathrm{i}\mathbf{k}\cdot\mathbf{r}} \chi_s \eta_t \tag{4.28}$$

and in the Dirac notation, it is marked as $|\nu\rangle$, where

$$|\nu\rangle = |n, s, t\rangle \equiv \varphi_\nu\,. \tag{4.29}$$

Because single-particle energy e_ν can correspond to two spin values ($\uparrow\downarrow$) and two isospin projections (p,n), so one is dealing here with a fourfold degeneration of the ν state, as shown in Fig. 4.4.

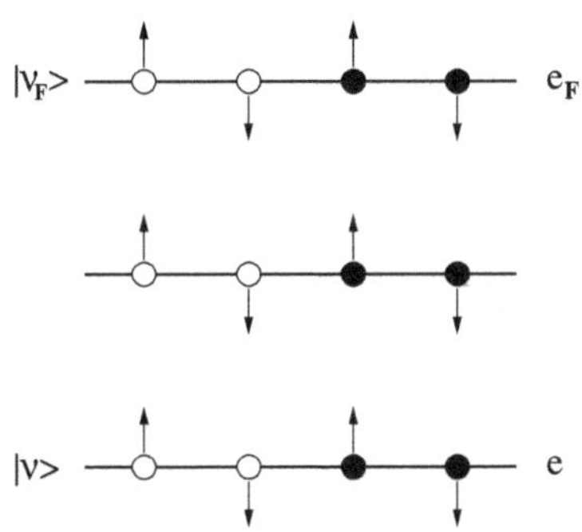

Figure 4.4 Occupation of states in the Fermi gas by neutrons (open symbol) and protons (filled circles) with spin up and down (arrows).

The average number of single-particle states of both types in the considered phase-space element is therefore

$$dN{=}4\,dn_1\,dn_2\,dn_3{=}4\frac{L}{2\pi}dk_1\,\frac{L}{2\pi}dk_2\,\frac{L}{2\pi}dk_3{=}4\left(\frac{L}{2\pi}\right)^3 d^3k{=}\frac{\mathscr{V}\,d^3k}{2\pi^3}\,. \tag{4.30}$$

One has assumed here that quantum numbers change continuously.
For the Fermi gas consisting of particles of one kind, protons (p) or neutrons (n), one has

$$dN^{p(n)} = \frac{\mathscr{V}\,d^3k}{4\pi^3}\,. \tag{4.31}$$

Let us assume that the highest occupied state $|\nu_{\mathrm{F}}\rangle$ for a given number N has energy e_{F}. The Fermi energy e_{F} and the wave number k_{F} are determined by the condition that the number of particles of one type below the Fermi surface is N. To find the number of nucleons in the states having k smaller than k_{F}, one must pass to the momentum coordinates and integrate the number of

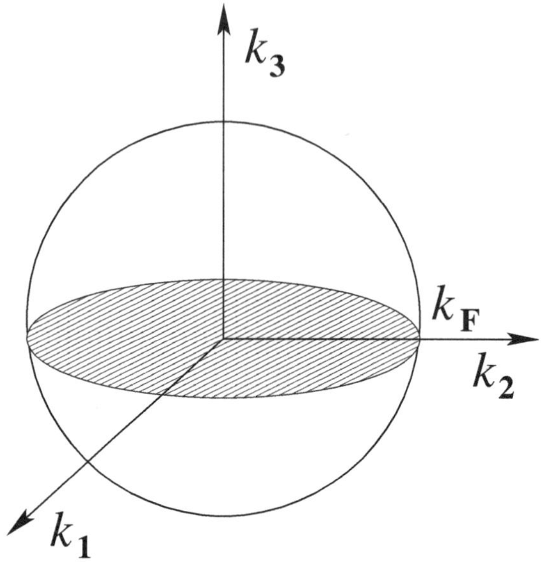

Figure 4.5 Momentum sphere of the Fermi gas.

particles over the volume of a sphere with radius $|k_F|$, as shown in Fig. 4.5. Hence, the number of one type particle is

$$N^{p(n)} = \int dN^{p(n)}, \tag{4.32}$$

where $dN^{p(n)}$ is specified in Eq. (4.31).

Then, one replaces the integration over discrete variable dN with integration over the continuous variable d^3k

$$N = \int_0^{2\pi}\int_{-1}^{1}\int_0^{k_F} \frac{\mathcal{V}}{4\pi^3} d^3k, \tag{4.33}$$

where

$$d^3k = d\varphi\, d\cos\vartheta\, k^2 dk.$$

The integral is

$$N = \frac{\mathcal{V} k_F^3}{3\pi^2}. \tag{4.34}$$

Hence

$$k_F^3 = \left(\frac{3\pi^2 N}{\mathcal{V}}\right) \tag{4.35}$$

and

$$e_{\mathrm{F}} = \left(\frac{3\pi^2 N}{\mathscr{V}}\right)^{2/3} \frac{\hbar^2}{2M}. \tag{4.36}$$

Inserting the density of nuclear matter $\rho_0 = (N+Z)/\mathscr{V} = 0.17$ nucleon/fm^3 into Eqs. (4.35)- (4.36) one gets for $N = Z$ the following estimate of the Fermi momentum and the Fermi energy

$$k_{\mathrm{F}} = 1.37\,\mathrm{fm}^{-1} \quad \text{and} \quad e_{\mathrm{F}} = 39\ \mathrm{MeV}.$$

The kinetic energy of the whole system will be obtained by integrating the energy element

$$dE = \frac{\hbar^2 k^2}{2M} dN^{p(n)} = \frac{\hbar^2 k^2}{2M} \frac{\mathscr{V}}{4\pi^3} d^3k \tag{4.37}$$

over a sphere with a radius of k_{F}

$$E = \int\int\int \frac{\hbar^2 k^2}{2M} \frac{\mathscr{V} d^3k}{4\pi^3} = \frac{\hbar^2}{2M} \frac{\Omega}{4\pi^3} \int_{-1}^{1} d\cos\vartheta \int_{0}^{2\pi} d\varphi \int_{0}^{k_{\mathrm{F}}} k^4 dk. \tag{4.38}$$

Hence

$$E = \frac{\hbar^2 \mathscr{V} k_{\mathrm{F}}^5}{10\pi^2 M}.$$

Using the relation (4.36), one can express full kinetic energy of gas by the Fermi energy

$$E = \frac{3}{5} N e_{\mathrm{F}}, \tag{4.39}$$

which corresponds to the average kinetic energy of a single-particle

$$\frac{E}{N} \simeq 23\ \mathrm{MeV}.$$

The Fermi gas model also allows to estimate the depth of the rectangular well's single-particle potential (V_0), as shown in Fig. 4.6. This depth is the sum of the nucleon separation energy ($S_{\mathrm{p(n)}}$) of about 8 MeV in heavy nuclei (Fig. 1.2) and the Fermi energy value

$$-V_0 = S_{\mathrm{p(n)}} + e_{\mathrm{F}} = (8+39)\ \mathrm{MeV} = 47\ \mathrm{MeV}.$$

In the Fermi gas model for protons and neutrons, the ground state energy of nucleus E_0 is

$$E_0 = \frac{3}{5}\left(N e_{\mathrm{F}}^{n} + Z e_{\mathrm{F}}^{p}\right). \tag{4.40}$$

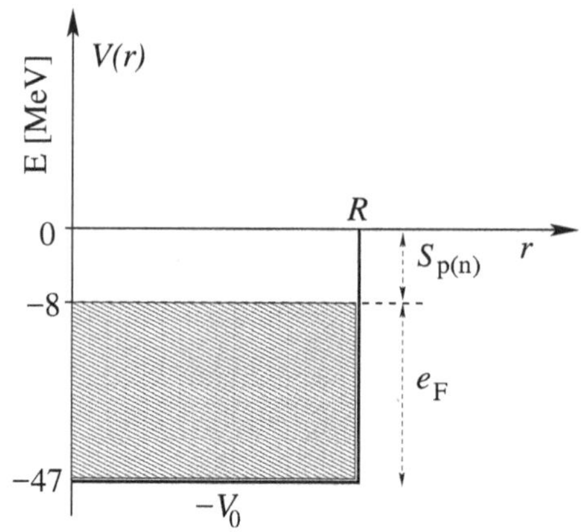

Figure 4.6 Estimate of the depth of the rectangular well of the average single-particle potential.

It can be represented in the form

$$E_0 = \frac{3}{5}\frac{\hbar^2}{2M}\left(\frac{3\pi^2}{\mathcal{V}}\right)^{2/3}\left(N^{5/3}+Z^{5/3}\right). \tag{4.41}$$

when one uses Eq. (4.36).

4.2.2 Symmetry energy

The relation (4.41) can be used to derive the symmetry term in Fermi gas energy, i.e., energy dependence on the neutron excess. A similar term in nuclear binding energy is in the liquid drop model. For this purpose, let us define a function

$$f(t) \equiv N^{5/3} + Z^{5/3} = \left(\frac{A+t}{2}\right)^{5/3} + \left(\frac{A-t}{2}\right)^{5/3}, \tag{4.42}$$

where $t \equiv N - Z$ and $A = N + Z$.
Expanding the function $f(t)$ into Taylor series around $N = Z = A/2$, i.e. $t = 0$

$$f(t) = f(0) + \frac{\partial f}{\partial t}\,|_{t=0}\, t + \frac{1}{2}\frac{\partial^2 f}{\partial t^2}\,|_{t=0}\, t^2 + \dots, \tag{4.43}$$

one gets

$$\begin{aligned} f(t) &= 2^{-5/3}\left[(A+t)^{5/3} + (A-t)^{5/3}\right] = \\ &= 2^{-5/3}\left[2A^{5/3} + \tfrac{10}{9}A^{-1/3}t^2\right]\dots. \end{aligned} \tag{4.44}$$

The energy of the system in its ground state (4.41) is therefore

$$E_0=\frac{3}{5}\left(\frac{\hbar^2}{2M}\right)\left(\frac{3\pi^2}{\mathcal{V}}\frac{A}{2}\right)^{2/3}\left(A+\frac{5}{9}\frac{(N-Z)^2}{A}\right). \tag{4.45}$$

Hence, by introducing Fermi energy in a nucleus of equal numbers of protons and neutrons

$$\mathrm{e}_{\mathrm{F}}^{N=Z}=\frac{\hbar^2}{2M}\left(\frac{3\pi^2 A}{2\mathcal{V}}\right)^{2/3},$$

one obtains

$$E_0=\frac{3}{5}\mathrm{e}_{\mathrm{F}}^{N=Z}A+\frac{1}{3}\mathrm{e}_{\mathrm{F}}^{N=Z}\frac{(N-Z)^2}{A}. \tag{4.46}$$

The energy E_0 consists of a volume term, $E_{\mathrm{vol}}=a_{\mathrm{vol}}A$ proportional to A and the symmetry term

$$E_{\mathrm{sym}}=a_{\mathrm{sym}}(N-Z)^2/A \tag{4.47}$$

dependent on the surplus neutrons.

One can estimate the coefficients of these terms, remembering that they only come from kinetic nucleon energy

$$a_{\mathrm{vol}}=\frac{3}{5}e_{\mathrm{F}}\approx 23\mathrm{MeV}, \tag{4.48}$$

$$a_{\mathrm{sym}}=\frac{1}{3}e_{\mathrm{F}}\approx 13\mathrm{MeV}. \tag{4.49}$$

Assuming equipartition between the kinetic and potential parts of symmetry energy, a total contribution of 26 MeV can be expected from symmetry energy to the total ground-state energy of the nucleus.

4.2.3 Average distance between single-particle levels

Using the Fermi gas model, one can also estimate average energy distances between single-particle levels in the nucleus. Let us calculate the energy difference of the systems with the number of particles (one type) $N+1$ and N. This ΔE energy difference is

$$\begin{aligned}\Delta E &= e_{\mathrm{F}}(N+1)-e_{\mathrm{F}}(N)=\left(\tfrac{3\pi^2}{\Omega}\right)^{2/3}\tfrac{\hbar^2}{2M}\left[(N+1)^{2/3}-N^{2/3}\right]=\\ &= e_{\mathrm{F}}(N)\left[\left(1+\tfrac{1}{N}\right)^{2/3}-1\right]\end{aligned} \tag{4.50}$$

and is the average distance between single-particle levels near Fermi level. Expanding the expression $(1+\epsilon)^{2/3}$ into Taylor's series, where $\epsilon = \frac{1}{N}$ is small for heavy nuclei, one gets

$$\Delta E = \frac{2}{3}\frac{e_{\mathrm{F}}}{N} \approx \frac{26\ \mathrm{MeV}}{N}, \tag{4.51}$$

which gives for the heaviest nuclei the distance $\Delta E \approx 200$ keV. This estimate is not far from experimental data.

CHAPTER 5

Macro-micro method

The backgrounds of the macroscopic-microscopic model, called in short 'macro-micro' method, were given by Myers and Swiatecki in the 60-es last century. Let us recall here the main ideas behind this model and its historical development. Following the way of thinking of Myers and Swiatecki will help the reader better understand the macroscopic models and the origin of the shell effects in nuclei. Modern versions of the macroscopic-microscopic model but with more exactly evaluated shell correction are still frequently used nowadays.

5.1 LIQUID DROP MODEL

Let us now present the historical development of the macroscopic model of nuclei, which assumes strong two-body interactions. This assumption is opposed to the Fermi gas model discussed in the previous chapter, in which nucleons do not interact with each other.

Atomic nuclei exhibit so-called mass defect

$$\Delta M(N,Z) = N\,M_{\rm n} + Z\,M_{\rm H} - M(N,Z)\,, \tag{5.1}$$

where $M_{\rm n}$ and $M_{\rm H}$ are the mass of the neutron and Hydrogen atom, respectively, and $M(N,Z)$ is the mass of the corresponding atom. Note that $\Delta M > 0$ because the sum of N neutron and Z Hydrogen atom masses is greater than the atom's mass. The mass defect defines the binding energy of the system :

$$B(N,Z) = \Delta M(N,Z)c^2\,. \tag{5.2}$$

Analyzing the plot of binding energy per one nucleon in the nucleus shown in Fig. 1.2 one can state that the maximum value of B/A is 8.8 MeV at $A \approx 70$. For the heaviest known nuclei, at $A \approx 250$, the binding energy is about 7.5

MeV. Despite the large value of the binding energy of heavy nuclei resulting from nuclear saturation, the existing difference of nuclear energy between $A \approx 100$ and $A \approx 240$ of 1 MeV per nucleon can be released in the process of fission of heavy nuclei into lighter fragments. The energy released in the fission process corresponds approximately to 1/1000 of the mass of the nucleus.

Let us estimate the energy obtained from the fission of 1kg of uranium

$$E = 10^{-3}\,\text{kg} \cdot (3 \cdot 10^{8}\frac{m}{s})^{2} = 9 \cdot 10^{13}\,\text{J}.$$

Assuming that within one day to heat one apartment, about $2 \cdot 10^{7}$ J is needed, the energy obtained from fission of this amount of uranium (^{235}U isotope) is sufficient to heat 12,500 flats annually.

The binding energy of atomic nuclei was first described in the year 1935 using a semi-empirical Bethe-Weizsaecker mass formula for a charged drop of nuclear liquid with the volume, surface, and Coulomb terms. In 1953, Green and Bethe proposed a more advanced formula for binding energy:

$$B = b_{\text{vol}}A - b_{\text{surf}}A^{2/3} - \frac{1}{2}b_{\text{sym}}\frac{(N-Z)^2}{A} - \frac{3}{5}\frac{Z^2e^2}{r_0}A^{-1/3} \equiv -E_{\text{LD}}\,. \tag{5.3}$$

Here, the volume term is linear in A, the surface energy is proportional to ($\sim A^{2/3}$), while the new symmetry term is proportional to the neutron excess $N-Z$ square. The Coulomb term corresponds to the electrostatic energy of a uniformly charged sphere, i.e., it is proportional to $Z^2/A^{1/3}$. The parameters of the above liquid drop (LD) formula:

$$\begin{aligned} b_{\text{vol}} &= 15.56\,\text{MeV}\,, \\ b_{\text{surf}} &= 17.23\,\text{MeV}\,, \\ b_{\text{sym}} &= 46.57\,\text{MeV}\,, \\ r_0 =& \;\; 1.24\,\text{fm}\,, \end{aligned} \tag{5.4}$$

were fitted to the experimentally measured masses.
The volume term reflects the constant binding energy per nucleon in the nuclear matter. The surface energy reduces the binding energy of the nucleus, reflecting that nucleons at the surface have fewer neighbors than those inside nuclei, i.e., they are weaker bound. The symmetry term reduces the binding energy for nuclei with growing difference N and Z. The Coulomb term grows rapidly with the increasing Z and reduces the binding energy significantly for heavy nuclei.

5.2 β STABILITY LINE

Several conclusions concerning nuclear stability can be derived from the liquid drop formula.

Let us write the formula for the mass of the nucleus, inserting the binding energy (5.3) to Eq. (5.1):

$$M(Z,N) = ZM_p + NM_n - b_{vol}A + b_{surf}A^{2/3} + b_{sym}\frac{(N-Z)^2}{2A} + \frac{3}{5}\frac{Z^2e^2}{r_0A^{1/3}}. \quad (5.5)$$

The nuclei, stable with respect to the β decay, have the smallest mass for a given value A. It can be written as

$$\frac{\partial M}{\partial N}|_{A=\mathrm{const}} = 0 \quad \text{or} \quad \frac{\partial M}{\partial Z}|_{A=\mathrm{const}} = 0. \quad (5.6)$$

The last condition takes the form

$$\frac{\partial}{\partial Z}\left[ZM_p + (A-Z)M_n - b_{vol}A + b_{surf}A^{2/3} + b_{sym}\frac{(A-2Z)^2}{2A} + \frac{3}{5}\frac{Z^2e^2}{r_0A^{1/3}}\right]_{A=\mathrm{const}} = 0, \quad (5.7)$$

hence

$$M_p - M_n - b_{sym}\frac{2(A-2Z)}{A} + \frac{6}{5}\frac{Ze^2}{r_0A^{1/3}} = 0. \quad (5.8)$$

Using the above formula, one can evaluate Z_0, which corresponds to the minimum of the binding energy for a given A.
Multiplying Eq. (5.8) by A

$$A(M_p - M_n) - 4b_{sym}(\frac{A}{2} - Z_0) + \frac{6}{5}\frac{e^2}{r_0}Z_0A^{2/3} = 0, \quad (5.9)$$

one obtains

$$A(M_n - M_p + 2b_{sym}) = Z_0\left(4b_{sym} + \frac{6}{5}\frac{e^2}{r_0}A^{2/3}\right). \quad (5.10)$$

From here, one can evaluate

$$Z_0 = \frac{A(M_n - M_p + 2b_{sym})}{4b_{sym} + \frac{6}{5}\frac{e^2}{r}A^{2/3}}. \quad (5.11)$$

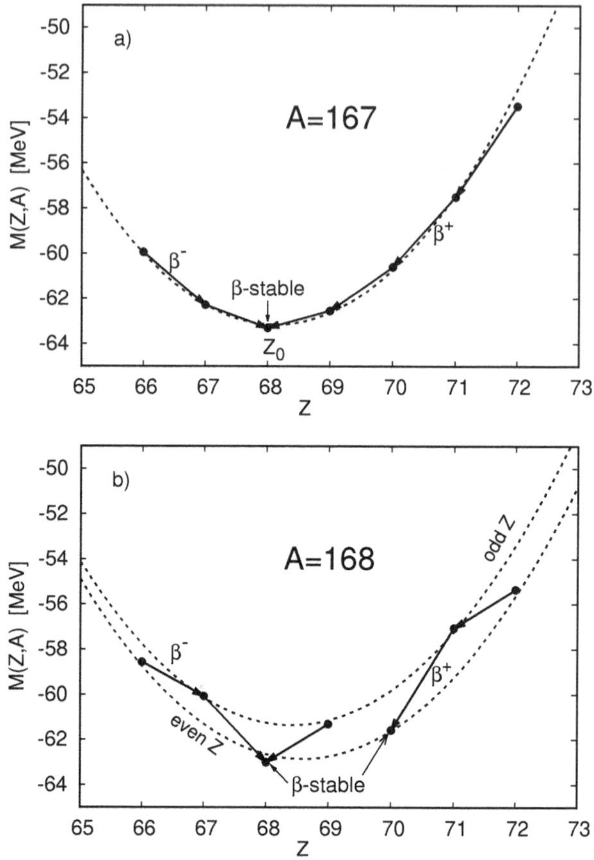

Figure 5.1 Masses of odd A=167 (a) and even A=168 (b) isobars as a functi ton number Z. The β^+ and β^- transitions are marked by arrows corresponding

It is the formula for the charge number Z_0 of the stable nucleus to neighboring isobaric isotopes decay, as shown in Fig. 5.1. This is the tion for the so-called β stability line, shown in Fig. 5.2, by a solid lir empirical formula given by Green can well approximate the β stability

$$(N-Z)_{\beta\text{stable}} = \frac{0.4 \cdot A^2}{A+200}.$$

The dashed lines in Fig. 5.2, called the neutron drip line or proton dri are the limits of bound neutrons or protons in the given nuclei. Th separation energies S_n of the neutron and the proton S_p correspond

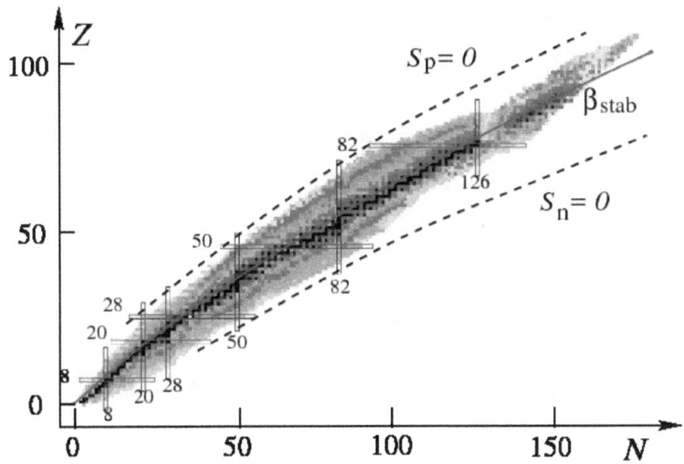

Figure 5.2 Peninsula of known nuclei in the (N, Z) plane. The black squares mark β-stable nuclei. The magic numbers for protons and neutrons are marked by the horizontal and vertical lines, respectively.

drip lines of the nuclear chart:

$$S_{\rm n}(N,Z) = B(N,Z) - B(N-1,Z)\,,$$
$$S_{\rm p}(N,Z) = B(N,Z) - B(N,Z-1)\,. \tag{5.13}$$

The equation for the neutron (or proton) separation energy equal to zero can be obtained by taking the derivative of binding energy over N (or Z)

$$S_{\rm n}(N,Z) \simeq \frac{\partial B}{\partial N}\,|_{Z=\rm const} = 0\,,$$
$$S_{\rm p}(N,Z) \simeq \frac{\partial B}{\partial Z}\,|_{N=\rm const} = 0\,. \tag{5.14}$$

Nuclei existing in nature and those that can be produced artificially in nuclear reactions are placed between these lines. Accurate prediction of the lines $S_{\rm p} = 0$ and $S_{\rm n} = 0$ is one of the most important tasks of theoretical models.

5.3 ENERGY OF PAIR CREATION

Formula (5.3) for nuclear binding energy does not contain a term that takes into account the experimentally observed enhancement of binding energy of even-even nuclei. The energy of creating pairs of the same kind of nucleons but with opposite spins can be estimated from the mass differences of neighboring even and odd nuclei, i.e., by the separation energy.

Consider a nucleus composed of an even-even number of particles core with energy $E_{\rm core}$ and a pair of nucleons on the outer shell. The energy of such a nucleus is

$$E = E_{\rm core} + E_{\rm int} + 2E_{\rm sp}\,, \tag{5.15}$$

where $E_{\rm sp}$ are the single-particle energies of valence nucleons and $E_{\rm int}$ is the energy of a particle pair bound by the pairing interaction. This energy half belongs to one nucleon and gives a contribution to binding energy

$$\Delta = E_{\rm int}/2\,, \tag{5.16}$$

which is called the pairing energy gap. The separation energy of N-th neutron is

$$S_{\rm n}(N,Z) = E_{\rm int} + E_{\rm sp}\,, \tag{5.17}$$

and that of the $N-1$ unbounded in a pair neutron is

$$S_{\rm n}(Z,N-1) = E_{\rm sp}\,. \tag{5.18}$$

Also

$$S_{\rm n}(Z,N+1) = E_{\rm sp}\,, \tag{5.19}$$

hence, on average, one has

$$E_{\rm sp} = \frac{1}{2}\{S_{\rm n}(N+1,Z) + S_{\rm n}(N-1,Z)\}\,. \tag{5.20}$$

So, the binding energy of a pair of neutrons is

$$E_{\rm int} = S_{\rm n}(N,Z) - \frac{1}{2}\{S_{\rm n}(N+1,Z) + S_{\rm n}(N-1,Z)\}\,, \tag{5.21}$$

or

$$\Delta = \frac{1}{4}[2S_{\rm n}(N,Z) - S_{\rm n}(N+1,Z) - S_{\rm n}(N-1,Z)]\,, \tag{5.22}$$

which, after inserting the separation energy definition (5.13), gives for neutrons

$$\Delta_{\rm n} = \frac{1}{4}\{B(N-2,Z) - 3B(N-1,Z) + 3B(N,Z) - B(N+1,Z)\}\,. \tag{5.23}$$

A similar equation one obtains for protons:

$$\Delta_{\rm p} = \frac{1}{4}\{B(N,Z-2) - 3B(N,Z-1) + 3B(N,Z) - B(N,Z+1)\}\,. \tag{5.24}$$

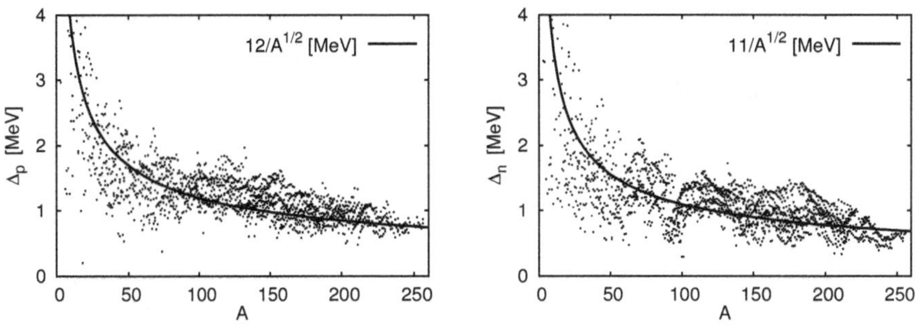

Figure 5.3 Experimental energy gaps (points) for protons Δ_p (l.h.s)) and neutrons Δ_n (r.h.s.) as function of the mass number A.

The energy gaps Δ_n and Δ_p deduced from the experimental masses are presented in Fig. 5.3 as a function of the mass number of nuclei. It turns out that for all mass numbers A the relationship $\Delta_p(Z)$ and $\Delta_n(N)$ can be approximated by the following functions

$$\Delta_p = \frac{12}{\sqrt{A}}\mathrm{MeV}, \qquad \Delta_n = \frac{11}{\sqrt{A}}\mathrm{MeV}. \tag{5.25}$$

The formula for the mass of the nucleus (5.5) should therefore be supplemented with the pairing term, responsible for the energy of pairs formation:

$$\delta E_{\mathrm{pair}} = \begin{cases} \Delta & \text{for even-even nuclei} \\ 0 & \text{for odd-even nuclei} \\ -\Delta & \text{for odd-odd nuclei} \end{cases}$$

and shell effects described in the next section

$$\begin{aligned} M(Z,N) &= ZM_p + NM_n - b_{\mathrm{vol}}A + b_{\mathrm{surf}}A^{2/3} + \\ &+ \frac{1}{2}b_{\mathrm{sym}}\frac{(N-Z)^2}{A} + b_{\mathrm{Coul}}\frac{Z^2}{A^{1/3}} + \delta E_{\mathrm{pair}} + \delta E_{\mathrm{shell}}, \end{aligned} \tag{5.26}$$

where $b_{\mathrm{Coul}} = \frac{3}{5}\frac{e^2}{r_0}$.

5.4 MYERS-SWIATECKI SHELL CORRECTION

Experimental data show that nuclei with the so-called magic numbers of protons or neutrons equal to 2, 8, 20, 28, 50, 82, and 126 for neutrons are particularly strongly bonded. The macroscopic formulas on the energy, like the liquid drop model, do not predict this effect. The idea of Myers and Swiatecki was to describe the binding energy of nuclei using two, for the first sight, contradictory models: macroscopic and microscopic. It was achieved by adding to the liquid drop energy $E_{\rm LD}$ the microscopic energy correction called the shell energy correction $E_{\rm shell}$:

$$E = E_{\rm LD} + E_{\rm shell} \,. \tag{5.27}$$

This correction comes from the quantum shell effects in the nucleus and is defined as the difference between the sum of single-particle energies e_ν of occupied levels and the energy of the nucleus without the shell structure $\tilde{E}$, i.e., with smoothed single-particle $\bar{e}_\nu$ spectrum

$$E_{\rm shell} = \sum_{\nu=1}^{A} e_\nu - \tilde{E} \,. \tag{5.28}$$

In general, the energies e_ν the single-particle levels $|\nu\rangle$ which are determined by solving the Schroedinger equation with some average single-particle potential. Below, the method of evaluating the shell correction energy proposed in 1965 by Myers and Swiatecki will be presented. Even though this method has only a historical significance, it is very instructive to follow its author's way of thinking. The Myers and Swiatecki shell correction was derived using the Fermi gas model, where from the continuous energy spectrum $E(n)$ one artificially creates a discrete spectrum, grouping single-particle states into shells corresponding to magic numbers as shown in Fig. 5.4. Calculations are made separately for protons and neutrons.

In the Fermi gas model the Fermi energy (4.36) for N particles of one type is:

$$e_{\rm F} = \frac{\hbar^2}{2M}\left(\frac{3\pi^2 N}{\mathcal{V}}\right)^{2/3} \,. \tag{5.29}$$

The energy of the level occupied by the n-th particle can be estimated by an analogous formula

$$e(n) = \frac{\hbar^2}{2M}\left(\frac{3\pi^2 n}{\mathcal{V}}\right)^{2/3} = \frac{\hbar^2}{2M}\left(\frac{3\pi^2 N}{\mathcal{V}}\right)^{2/3}\left(\frac{n}{N}\right)^{2/3} = e_{\rm F}\left(\frac{n}{N}\right)^{2/3} \,. \tag{5.30}$$

Let's assume that it will be the average single-particle energy in the nucleus with a smoothed shell structure

$$\bar{e} = e(n) = \frac{e_{\rm F}}{N^{2/3}} \cdot n^{2/3} \,. \tag{5.31}$$

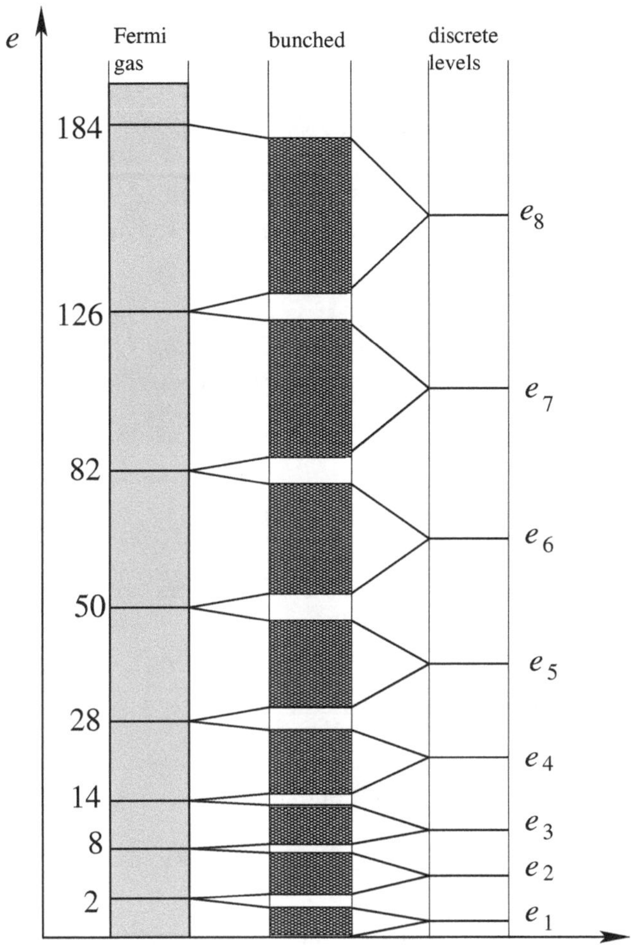

Figure 5.4 Myers' and Swiatecki's idea of bunching the single-particle levels.

Then, they divided the continuous energy spectrum into partitions corresponding to magic numbers N_i, whose ranges are: $N_i - N_{i-1} = 2, 6, 12, 8, 22, 32, 44, \ldots$ for $i = 1, 2, 3, \ldots$. Total nucleon energy in the i-th partition is

$$
\begin{aligned}
E_i &= \int_{N_i-1}^{N_i} e(n)dn = \frac{e_F}{N^{2/3}} \int_{N_i-1}^{N_i} n^{2/3} dn \\
&= \frac{3}{5} \frac{e_F}{N^{2/3}} \cdot \left[N_i^{5/3} - (N_{i-1})^{5/3} \right], \qquad (5.32)
\end{aligned}
$$

and the average energy per nucleon in the i-th range is

$$e_i = \frac{E_i}{N_i - N_{i-1}} = \frac{3}{5}\frac{e_{\rm F}}{N^{2/3}}\left(\frac{N_i^{5/3} - N_{i-1}^{5/3}}{N_i - N_{i-1}}\right). \tag{5.33}$$

Myers and Swiatecki took these energies as the position of single-particle levels in a real nucleus with a shell structure. When one replaces sums with integrals, the shell correction will be

$$E_{\rm shell} = \int_0^N (e_i - \bar{e})dn = \int_0^N \frac{e_{\rm F}}{N^{2/3}}\left[\frac{3}{5}\left(\frac{N_i^{5/3} - N_{i-1}^{5/3}}{N_i - N_{i-1}}\right) - n^{2/3}\right]dn. \tag{5.34}$$

It is convenient to express the shell correction using a function

$$F(N) \equiv \int_0^{\rm N} \left[q(n) - n^{2/3}\right], \tag{5.35}$$

where $q(n)$ is a step function shown in Fig. 5.5 with successive values determined in the 'centers of gravity' of the shells

$$q_i(N) \equiv \frac{3}{5}\left(\frac{N_i^{5/3} - N_{i-1}^{5/3}}{N_i - N_{i-1}}\right). \tag{5.36}$$

After performing integrating in Eq. (5.35) one obtains

$$F(N) = q_i(N)(N - N_{i-1}) - \frac{3}{5}\left(N^{5/2} - N_{i-1}^{5/2}\right), \tag{5.37}$$

where $N_{i-1} < N < N_i$, because only the last interval gives a non-zero contribution to the integral.

The process of averaging the step function around the certain curve $n^{2/3}$ is followed by the phenomenon of so-called 'flattering' of this function. One has to lower q_i about $cN/2^{2/3}$ as the functions $n^{2/3}$ and $q(n)$ should go along each other to make the shell correction oscillating around zero. A shell correction for the whole nucleus is the sum of corrections derived from neutrons and protons (see 5.34)

$$E_{\rm shell} = \frac{e_{\rm F}^{\rm n}}{N^{2/3}}\left(F(N) - \frac{cN}{2^{2/3}}\right) + \frac{e_{\rm F}^{\rm p}}{Z^{2/3}}\left(F(Z) - \frac{cZ}{2^{2/3}}\right). \tag{5.38}$$

Assuming that $e_{\rm F}^{\rm p} \approx e_{\rm F}^{\rm n} = e_{\rm F}$ is valid for nuclei close to β-stability, one obtains:

$$E_{\rm shell} = \frac{e_{\rm F}}{(A/2)^{2/3}}\left\{F(N) + F(Z) - \frac{cA}{2^{2/3}}\right\}. \tag{5.39}$$

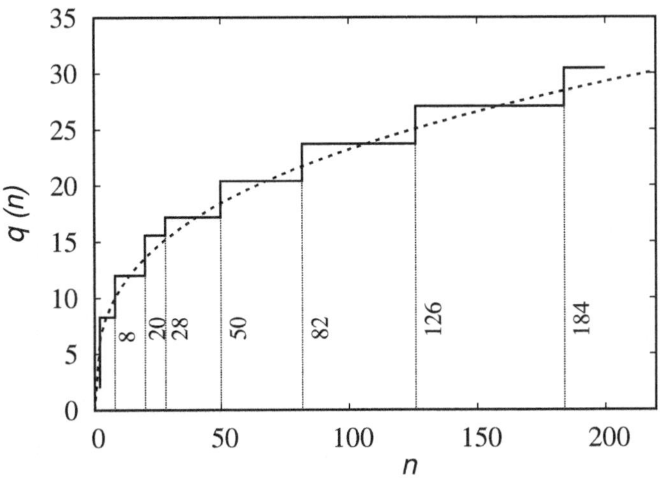

Figure 5.5 Single-particle energies $q(n)$ in the discrete and degenerated spectrum by Myers-Swiatecki (solid line) as a function of particle numbers compared to the Fermi gas energies $n^{2/3}$ (dotted line).

Hence, the Myers and Swiatecki shell correction can be written in the following form:

$$E_{\text{shell}} = e_{\text{F}} \left\{ \frac{F(N) + F(Z)}{(A/2)^{2/3}} - cA^{1/3} \right\}. \tag{5.40}$$

Fig. 5.6 shows the function $E_{\text{shell}}(N,Z)$, which increases binding energy for nuclei with a magical number of protons or neutrons and reduces the binding energy of systems with unfilled shells.

5.5 DEFORMATION DEPENDENCE OF BINDING ENERGY

The next step in developing the liquid drop model was considering the nuclear binding energy dependence on the nucleus's shape. Only a few terms, like the surface and Coulomb energies, vary with the nucleus's deformation in the liquid drop formula. The volume term does not change with deformation as the nucleus volume remains constant due to the high incompressibility of the nuclear matter. Also, the shell correction should depend on the shape of the nucleus.

Let us introduce the function being the ratio of the deformed nucleus surface $S\,(\text{def})$ to the surface of the $S\,(0)$ sphere of the same volume

$$B_s(\text{def}) = \frac{S\,(\text{def})}{\text{S}(0)} \tag{5.41}$$

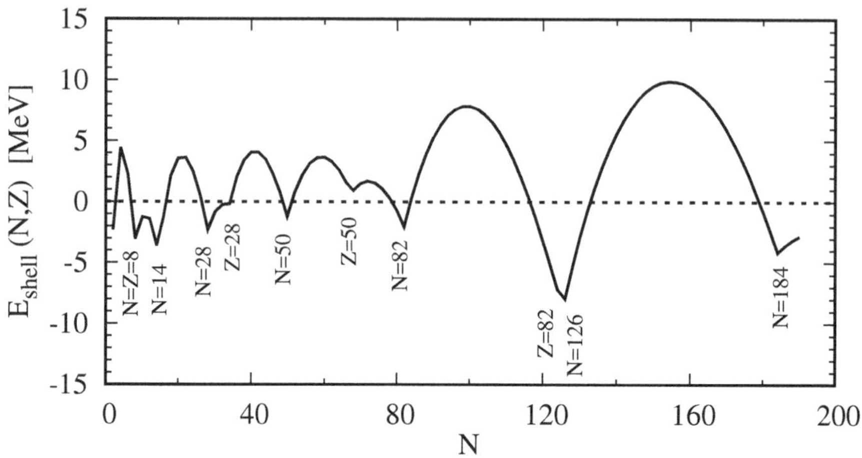

Figure 5.6 Myers-Swiatecki shell correction energy $E_{\text{shell}}(N, Z)$ for nuclei along the β-stability line as function of the neutron number.

and an analogous ratio of the Coulomb energies for uniformly charged drop

$$B_c(\text{def}) = \frac{E_{\text{Coul}}(\text{def})}{E_{\text{Coul}}(0)}. \tag{5.42}$$

The liquid drop formula (5.3) then takes the form

$$\begin{aligned} E_{\text{LD}}(Z,N,\text{def}) &= -b_{\text{vol}}A + b_{\text{surf}}A^{2/3}B_s(\text{def}) + \frac{1}{2}b_{\text{sym}}\frac{(N-Z)^2}{A} + \\ &+ \frac{3}{5}\frac{Z^2e^2}{r_0A^{1/3}}B_c(\text{def}). \end{aligned} \tag{5.43}$$

The problem of parameterization of the nucleus shape emerges here. Adding deformation parameters to the particle coordinates is always associated with an artificial increase in the degrees of freedom of nucleons in the nucleus. This effect is dangerous to obtain, e.g., spurious excited states in the nucleus. However, the convenience of macroscopic approaches to the shape description of nuclei has determined a high popularity of such models.

5.6 SHAPE PARAMETERIZATION OF NUCLEI

One can obtain a complete description of the shape by expanding the surface of the nucleus in an infinite series of spherical functions

$$R(\vartheta,\varphi) = R_0(\{a_{\lambda\mu}\})\left[1 + \sum_{\lambda=0}^{\infty}\sum_{\mu=-\lambda}^{\lambda} a_{\lambda\mu}Y_{\lambda\mu}(\vartheta,\varphi)\right], \tag{5.44}$$

where R_0 is obtained from the condition of the volume conservation of the deformed nucleus. The parameters $a_{\lambda\mu}$ describe various shapes, among which: λ even describe bodies that are reflection symmetric concerning z-axis, $\mu = 0$ - axially symmetric shapes, a_{20} - quadrupole deformed nuclei, a_{30} - octupole deformed, pear-like shapes, a_{40} - hexadecapole, etc.

Usually, the series (5.44) ends at the first few terms and some combinations of $a_{\lambda\mu}$ parameters are often created to make calculations simpler. Of course, the parameterization (5.44) is not the only one possible. In practical applications, other, more convenient descriptions of the deformed nucleus shape are often used (confer, e.g., Sec. 6.5). The transition from one parameterization to the other would be exact when an infinite number of the parameters $a_{\lambda\mu}$ is taken. One frequently uses simpler phenomenological formulas to describe specific shapes of deformed nuclei. The deformation parameters are entered in a macroscopic way, describing the shape of a drop of nuclear matter, or microscopically, by the shape of an average potential field. One will discuss microscopic parameters in the chapters about shell models, and now, only some examples of macroscopic parameters entered for the coefficients $a_{\lambda\mu}$ describing quadrupole-deformed nuclei will be presented.

Shapes close to axially symmetric spheroids are described by quadrupole deformation with $\lambda = 2$ and $\mu = 0$. The surface equation takes the form

$$R(\vartheta) = R_0(\beta)[1 + \beta Y_{20}(\vartheta)], \tag{5.45}$$

where $\beta \equiv a_{20}$ is a quadrupole deformation parameter. For non-axial case but with the mirror symmetry concerning each main axis, one has $\lambda = 2$ and $\mu = 0, \pm 2$, and the surface of the nucleus is described by

$$R(\vartheta,\varphi) = R_0(\{a_{2\mu}\})\left[1 + \sum_{\mu=-2,0,2} a_{2\mu}Y_{2\mu}(\vartheta,\varphi)\right]. \tag{5.46}$$

Then one can introduce the Bohr parameterization β,γ, where

$$\begin{aligned} a_{20} &= \beta\cos\gamma, \\ a_{22} = a_{2-2} &= \tfrac{1}{\sqrt{2}}\beta\sin\gamma. \end{aligned} \tag{5.47}$$

There is no term with $\mu = \pm 1$ in Eq. (5.47), because the axes of the coordinate system are the major axes of the ellipsoid. The global deformation of the nucleus could be characterized by the α parameter introduced by Myers and Swiatecki, which is a measure of the relative deviation of the deformed surface from the sphere

$$\alpha^2 = \int\int \frac{[R(\vartheta,\varphi) - R_0]^2 d\cos\vartheta\, d\varphi}{R_0^2} = \sum_{\lambda\mu} (a_{\lambda\mu})^2 . \tag{5.48}$$

In the case of quadrupole-deformed nuclei, one has

$$\alpha^2 = a_{20}^2 + a_{22}^2 + a_{2-2}^2 \equiv \beta^2 . \tag{5.49}$$

The shape-dependent functions B_s and B_c in the parameterization (5.46-5.49) read

$$\begin{aligned} B_s(\beta,\gamma) &= 1 + \frac{2}{5}\beta^2 - \frac{4}{105}\beta^3 \cos\gamma + 0(\beta^4), \\ B_c(\beta,\gamma) &= 1 - \frac{1}{5}\beta^2 - \frac{4}{105}\beta^3 \cos\gamma + 0(\beta^4). \end{aligned} \tag{5.50}$$

Making the liquid drop energy and the shell correction energy dependent on deformation, one can find the minimum of the total energy of the nucleus (i.e., the ground state) at a non-zero deformation point and evaluate the height of the potential barrier due to nuclear fission (i.e., large elongation). The condition for the existence of the LD fission barrier is that the surface energy with deformation has to grow faster than Coulomb energy decreases

$$b_{\rm surf} A^{2/3} \cdot \frac{2}{5}\beta^2 \geq b_{\rm Coul} \frac{Z^2}{A^{1/3}} \cdot \frac{\beta^2}{5} \tag{5.51}$$

or

$$\frac{b_{\rm Coul} Z^2}{2 b_{\rm surf} A} \leq 1 . \tag{5.52}$$

Let us define the fissility parameter

$$x \equiv \frac{b_{\rm Coul} Z^2}{2 b_{\rm surf} A} \approx \frac{Z^2}{49A}, \tag{5.53}$$

which describes in the liquid drop model the properties of the nuclei due to fission. For nuclei that do not fission, i.e., have got a high potential barrier due to fission, the parameter $x \ll 1$. For example, for non-fissioning lead isotope ^{208}Pb $x = 0.66$, on the other hand, for the spontaneously fissioning ^{240}Pu, the fissility parameter is already 0.75, and for the isotope of the super-heavy element ^{288}Fl, the parameter achieves value close to 1 (x =0.92). When $x > 1$, there is no LD barrier to fission.

Please note that the x parameter only describes the global macroscopic properties of nuclei. In specific cases, the fission barrier height depends significantly on the shell effects in the deformed nucleus. In the Myers and Swiatecki model, it was assumed that deformation destroys shell structure because it breaks the spherical symmetry of the nucleus, reducing the shell correction energy. The dependence of the shell correction on deformation is according to Myers and Swiatecki model given by

$$E_{\text{shell}}^{\text{def}}(Z,N,\alpha) = E_{\text{shell}}(Z,N)\exp\left(\frac{-\alpha^2}{\alpha_0^2}\right), \tag{5.54}$$

where α_0 is a phenomenological parameter matched to experimentally determined barrier heights. However, later studies have shown that a strong shell structure can also appear in deformed nuclei (shape isomers).

Nuclei with the number of nucleons far from the magic numbers have a positive shell correction, which creates a minimum at non-zero deformation α_{eq}. This effect is presented graphically in Fig. 5.7 (a). Nuclei with magic or close to magic nucleon numbers (b) have a higher fission barrier and are more stable because they have a negative shell correction. This correction, added to the liquid drop energy E_{LD} (5.43), deepens the spherical minimum of the total energy.

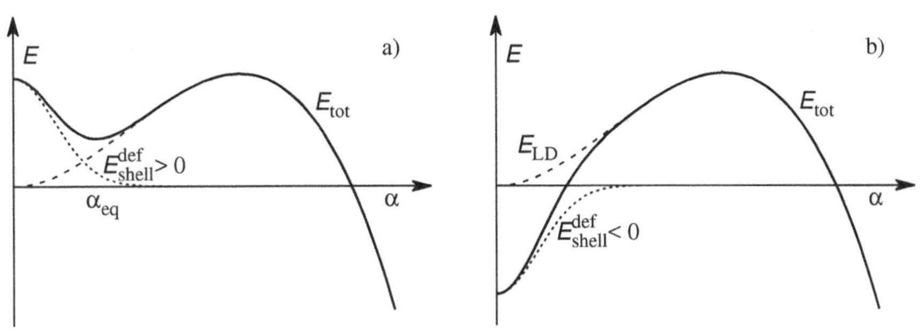

Figure 5.7 Liquid drop energy E_{LD} (dashed line) and shell correction $E_{\text{shell}}^{\text{def}}$ (dotted line) as a function of the deformation parameter α for nuclei with nucleons numbers far from the magic numbers (a) or near to them (b). The solid line shows the nucleus's total potential energy E_{tot}.

5.7 MASS FORMULA BY MYERS-SWIATECKI

Mass formula of Myers and Swiatecki is

$$\begin{aligned} M(Z,N,\alpha,\gamma) &= ZM_{\mathrm{p}}+NM_{\mathrm{n}}-C_1A \\ &+ C_2A^{2/3}\left(1+\frac{2}{5}\alpha^2-\frac{4}{105}\alpha^3\cos\gamma\right) \\ &+ C_3\frac{Z^2}{A^{1/3}}\left(1-\frac{1}{5}\alpha^2-\frac{4}{105}\alpha^3\cos\gamma\right) \\ &- C_4\frac{Z^2}{A}+\Delta E_{\mathrm{pair}}(Z,N)+E_{\mathrm{shell}}(Z,N)\mathrm{e}^{-(\alpha/\alpha_0)^2}. \end{aligned} \quad (5.55)$$

The C_4Z^2/A term that one has not discussed yet is a correction to the Coulomb energy of the nucleus due to the smooth surface charge density. In addition, both volume and surface terms contain symmetry energy

$$\begin{aligned} C_1 &= a_1\left[1-\kappa\left(\frac{N-Z}{A}\right)^2\right], \\ C_2 &= a_2\left[1-\kappa\left(\frac{N-Z}{A}\right)^2\right]. \end{aligned} \quad (5.56)$$

The pairing term, describing the energy of pairing, was adopted in the form

$$\Delta E_{\mathrm{pair}}(Z,N)=\begin{cases} \dfrac{11}{\sqrt{A}}\ \mathrm{MeV} & \text{for } Z, N \text{ odd}, \\ 0\ \mathrm{MeV} & \text{for } A \text{ odd}, \\ -\dfrac{11}{\sqrt{A}}\ \mathrm{MeV} & \text{for } Z,N \text{ even}. \end{cases} \quad (5.57)$$

The shell correction is given by the formula (5.40)

$$E_{\mathrm{shell}}(Z,N)=C\left(\frac{F(Z)+F(N)}{(A/2)^{2/3}}-cA^{1/3}\right). \quad (5.58)$$

The mass formula has nine freely adjusted parameters (5.55). They were fitted in 1965 by Myers and Swiatecki to 1200 masses of nuclei, 40 fission barriers, and 240 quadrupole moments known experimentally at that time. They have got the following set of parameters

$$\begin{array}{lll} a_1 = 15.4941\ \mathrm{MeV}, & a_2 = 17.9439\ \mathrm{MeV}, & C_3 = 0.7053\ \mathrm{MeV}, \\ r_0 = 1.2249\ \mathrm{fm}, & C_4 = 1.21129\ \mathrm{MeV}, & \kappa = 1.7826, \\ C = 5.8\ \mathrm{MeV}, & c = 0.26, & \alpha_0 = 0.27. \end{array} \quad (5.59)$$

In this simple phenomenological model, masses were predicted with the accuracy of 2 MeV, while without adding the shell and pairing corrections, the error reached even 12 MeV in some cases. The existence of deformed nuclei shown in Fig. 5.2 was also correctly predicted.

5.8 DROPLET MODEL

The extended version of the liquid drop model is the droplet model. It contains corrections due to the proton and neutron distribution density fluctuations and surface curvature of nuclei. It is obtained by decomposing the integral of the nucleus's energy density in power series $A^{1/3}$, assuming a thin fuzzy surface layer of the thickness a, where $a/R \ll 1$. R is the nucleus radius with a mass number A so large that $A^{-1/3} \ll 1$.

The following formula gives the quantum mechanical density of the nucleus:

$$\rho = A \int\int\int \cdots \int \Psi^\star \Psi d\tau_2 \cdots d\tau_A, \tag{5.60}$$

while the energy density functional is

$$\eta = \int\int \cdots \int \Psi^\star \hat{\mathrm{H}} \Psi d\tau_2 \cdots d\tau_A. \tag{5.61}$$

Here $\hat{H}$ is a many-body Hamiltonian, Ψ is the ground-state function of many nucleons, and τ_i are coordinates of the i-th particle. The integration takes place over the whole space.

Nuclear energy, being the volume integral from energy density (5.61), can be represented in a volume term and corrections in the form of an integral over the nucleus surface

$$\begin{aligned} B = \int_V \eta d^3r &= \int_V (\eta + b_{\text{vol}}\rho - b_{\text{vol}}\rho) d^3r = \\ &= b_{\text{vol}} \int_V \rho d^3r + \int_V (\eta - b_{\text{vol}}\rho) d^3r \\ &= b_{\text{vol}} A + \int_V (\eta - b_{\text{vol}}\rho) d^3r \\ &= b_{\text{vol}} A + \int_\Sigma d\sigma \int_0^\infty (\eta - b_{\text{vol}}\rho) dr. \end{aligned} \tag{5.62}$$

The radial integral in the last term is the surface tension and depends on local properties, e.g., surface curvature of nuclei

$$\gamma \equiv \int_0^\infty (\eta - b_{\rm vol}\rho)\, dr = \gamma(a\kappa, a^2\Gamma). \tag{5.63}$$

In Fig. 5.8 the nuclear density ρ and the integral function $\eta - b_{\rm vol}\rho$ from Eq. (5.62) are schematically shown.

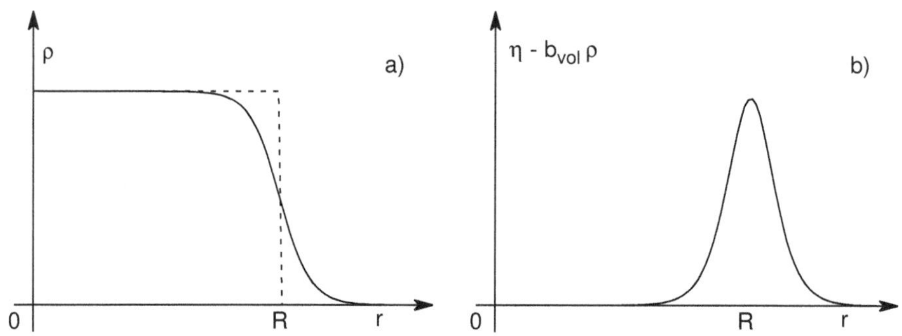

Figure 5.8 Nuclear density ρ (a) and the surface energy $\eta - b_{\rm vol}\rho$ (b) as a function of the distance from the center r.

The surface tension γ is a function of the average curvature κ and Gaussian curvature Γ, where

$$\kappa = \frac{1}{R_1} + \frac{1}{R_2}, \quad \Gamma = \frac{1}{R_1 R_2}, \tag{5.64}$$

R_1 and R_2 are the local main radii of the surface curvature, and the a parameter describes the nucleus surface diffuseness. Let us develop the functional γ (5.63) into the Taylor series around a sphere of the same volume as the deformed nucleus

$$\gamma(a\kappa, a^2\Gamma) = \gamma + \gamma'_\kappa \kappa a + \frac{1}{2}\gamma''_{\kappa\kappa}\kappa^2 a^2 + +\gamma'_\Gamma \Gamma a^2 + \cdots \tag{5.65}$$

and put it into the last integral of (5.62). Hence, the energy correction to the volume term is

$$\begin{aligned}\int_V (\eta - b_{\rm vol}\rho) d^3\mathbf{r} &= \gamma \int_\Sigma d\sigma + \gamma'_\kappa a \int_\Sigma \kappa d\sigma + \\ &+ \frac{1}{2}\gamma''_{\kappa\kappa} a^2 \int_\Sigma \kappa^2 d\sigma + \gamma'_\Gamma a^2 \int_\Sigma \Gamma d\sigma + \cdots .\end{aligned} \tag{5.66}$$

Individual terms in the last equation can be written as

$$\begin{aligned}&\gamma\int_{\Sigma} d\sigma \equiv -b_{\text{surf}}A^{2/3},\\&\gamma'_{\kappa}a\int_{\Sigma}\kappa d\sigma \equiv -b_{\text{curv}}A^{1/3},\\&\frac{1}{2}\gamma''_{\kappa\kappa}a^2\int_{\Sigma}\kappa^2 d\sigma+\gamma'_{\Gamma}a^2\int_{\Sigma}\Gamma d\sigma \equiv -b_0A^0,\end{aligned} \tag{5.67}$$

where an explicit dependence of each term on the number of nucleons A is written.

The nuclear binding energy (5.64) is therefore given by the equation

$$B = b_{\text{vol}}A - b_{\text{surf}}A^{2/3} - b_{\text{curv}}A^{1/3} - b_0A^0\cdots. \tag{5.68}$$

The following new elements appear in the nucleus's energy: the term dependent on $A^{1/3}$, responsible for the thin surface layer's curvature, and the one which does not depend on the mass number A. Of course, in the formula (5.68), one only considered contributions originating from the nuclear interaction, and the Coulomb energy is omitted, which should be subtracted from the nuclear part of the binding energy.

The droplet model minimizes nuclear binding energy relative to fluctuations in the density distribution of protons and neutrons and deviations of the nucleon distribution from the sharp nucleus surface. More extended calculations, which will not be quoted here, lead to the following expression for energy

$$\begin{aligned}B(N,Z,\text{def}) &= \left[-a_1+J\bar{\delta}^2+\frac{1}{2}K\bar{\varepsilon}^2+\frac{1}{2}M\bar{\delta}^4\right]A\\&+\left[a_2+\frac{9}{4}\frac{J^2}{Q}+\bar{\delta}^2\right]A^{2/3}B_{\text{s}}\\&+a_3A^{1/3}B_{\text{k}}+C_1Z^2A^{-1/3}B_{\text{c}}-C_2Z^2A^{1/3}B_{\text{r}}\\&-C_5Z^2B_{\text{w}}-C_3Z^2A^{-1}-C_42^{-1/3}Z,\end{aligned} \tag{5.69}$$

where

$$\begin{aligned}&\bar{\delta}=\left[I+\frac{3}{16}(C_1/Q)ZA^{-2/3}B_{\text{v}}\right]/\left[1+\frac{9}{4}J/QA^{-1/3}B_{\text{s}}\right],\\&\bar{\varepsilon}=\left[-2a_2A^{-1/3}B_{\text{s}}+L\bar{\delta}^2+C_1Z^2A^{-4/3}B_{\text{c}}\right]/K.\end{aligned} \tag{5.70}$$

The $I = \frac{N-Z}{A}$ parameter in the above formula is reduced isospin of nucleus. The factors B_s, B_k, B_c, B_r, B_v, and B_w, appearing in formula (5.69) are functionals dependent on the shape of the nucleus. They describe the relative change of the nucleus surface, curvature, Coulomb energy, and its spatial distribution related to the deviation of the Coulomb potential $W(r)$ from its mean value

$$\bar{W} = \int_V W(r)d\tau/(\frac{4}{3}\pi R^3) ,$$

both inside and on the surface of the nucleus. These coefficients depend on the shape of the nucleus, and they can be calculated by integrating over volume V, or the S nucleus surface:

$$\begin{aligned}
B_s &= \int_S d\sigma/(4\pi R^2), \\
B_c &= \int_V W(r)d\tau/\left(\frac{32}{15}\pi^2 R^5\right), \\
B_k &= \int_S \left(\frac{1}{R_1}+\frac{1}{R_2}\right)d\sigma/(8\pi R), \\
B_r &= \int_V [W(r)-\bar{W}]^2 d\tau/\left(\frac{64}{1575}\pi^2 R^7\right), \\
B_v &= \int_S [W(r)-\bar{W}]d\sigma/\left(-\frac{16}{15}\pi^2 R^4\right), \\
B_w &= \int_S [W(r)-\bar{W}]^2 d\sigma/\left(-\frac{64}{225}\pi^3 R^6\right).
\end{aligned} \tag{5.71}$$

All integrals in (5.71) obtained for the deformed nucleus are divided by integral values calculated analytically for the spherical nucleus so that the factors B_i are equal to one when the nucleus is not deformed.

The following set of parameters was obtained by Myers and Swiatecki by matching energy, masses, fission barrier heights, and quadrupole moments of the nuclei, calculated within the droplet model, to experimental data:

$$\begin{aligned}
J &= 36.5 \text{ MeV} , & M &= 0, \\
Q &= 17 \text{ MeV} , & r_0 &= 1.175 \text{ fm}, \\
K &= 240 \text{ MeV} , & b &= 0.99\text{fm}, \\
L &= 100 \text{ MeV} .
\end{aligned} \tag{5.72}$$

The coefficients C_i appearing in Eqs.(5.69)-(5.70) are the following functions of the above parameters:

$$
\begin{aligned}
C_1 &= \frac{3}{5}\frac{e^2}{r_0} = 0.73531\,, \\
C_2 &= \frac{C_1^2}{336}\left(\frac{1}{J}+\frac{18}{K}\right) = 0.00016477\,, \\
C_3 &= \frac{5}{2}C_1\left(\frac{b}{r_0}\right)^2 = 1.30501\,, \\
C_4 &= \frac{5}{4}C_1\left(\frac{3}{2\pi}\right)^{2/3} = 0.56149\,, \\
C_5 &= \frac{1}{64}\frac{C_1^2}{Q} = 0.00049695\,.
\end{aligned}
\tag{5.73}
$$

The droplet formula for nuclear binding energy (5.69) was in 80-es, the most commonly used macroscopic formula for calculating masses of spherical nuclei. Unfortunately, it does not reproduce well the nuclear energy dependence on deformation. The following section will present another phenomenological model that better describes this relationship based on the Yukawa interaction.

5.9 YUKAWA MACROSCOPIC ENERGY

The macro-micro model developed by Krappe and Nix in 1973 is based on the Yukawa forces (see Eq.2.4). It is often used in calculations of properties of strongly deformed nuclei. Folding the density distribution of nucleons in a nucleus with the Yukawa force gives an average mean-field potential acting on nucleons and the corresponding energy density functional. This model's advantage is that due to the folding, it offers a more consistent method of calculating the macroscopic part of nuclear potential energy and the energy correction describing quantum shell effects in the nucleus (5.28).

Considerations carried out for nuclear matter have shown that some modifications should be added to the Yukawa forces to get by this simple model the proper value of the binding energy of matter. This modification relies on addition of an exponential function to Yukawa's forces. However, the details of this model (known in the literature under the name "Yukawa plus exponential") will not be discussed here, and only the final formula for the mass of the nucleus is quoted below:

$$
\begin{aligned}
M_{\mathrm{Y}}(Z,N,\mathrm{def}) &= M_{\mathrm{H}}Z + M_{\mathrm{n}}N - a_{\mathrm{el}}Z^{2.39} - a_{\mathrm{v}}(1-\kappa_{\mathrm{v}}I^2)A \\
&+ a_{\mathrm{s}}(1-\kappa_{\mathrm{s}}I^2)A^{2/3}B_1 + c_0A^0 + c_1Z^2A^{-1/3}B_3 \\
&- c_4Z^{-4/3}A^{-1/3} + f(k_{\mathrm{F}}r_{\mathrm{p}})Z^2A^{-1} - c_{\mathrm{a}}(N-Z) \\
&+ W\left(|I| + \begin{cases} 1 & \text{for } Z=N \text{ odd} \\ 0 & \text{in all other cases} \end{cases}\right) \\
&+ \left(\begin{array}{rl} \Delta A^{-1/2} + \frac{1}{2}\delta & \text{for } Z \text{ and } N \text{ odd} \\ \frac{1}{2}\delta & \text{for } Z \text{ or } N \text{ odd} \\ -\Delta A^{-1/2} - \frac{1}{2}\delta & \text{for } Z \text{ and } N \text{ even} \end{array}\right).
\end{aligned}
\tag{5.74}
$$

Here M_{H}=7.289034 MeV and M_{n}=8.071431 MeV are the hydrogen atom and neutron masses, respectively, and x^2=1.4399764 MeV·fm is the square of electron charge. The Coulomb energy coefficients occurring in this formula are:

$$
c_1 = \frac{3}{5}\frac{e^2}{r_0}, \quad c_4 = c_1 \cdot \frac{5}{4}\left(\frac{3}{2\pi}\right)^{2/3}, \tag{5.75}
$$

and the proton form factor f is given by the following formula:

$$
f(k_{\mathrm{F}}r_{\mathrm{p}}) = -\frac{1}{8}\frac{r_{\mathrm{p}}^2e^2}{r_0^3}\left[\frac{145}{48} - \frac{327}{2880}(k_{\mathrm{F}}r_{\mathrm{p}})^2 + \frac{1527}{1209600}(k_{\mathrm{F}}r_{\mathrm{p}})^4\right], \tag{5.76}
$$

where k_{F} is the length of the Fermi wave-vector:

$$
k_{\mathrm{F}} = \left(\frac{9\pi Z}{4A}\right)^{1/3}\frac{1}{r_0}.
$$

B_1 in Eq. (5.74) is the surface energy for finite range forces

$$
B_1 = \frac{A^{-2/3}}{8\pi^2 r_0^2 a^4}\int_V\int_{V'}\left(2 - \frac{|\mathbf{r}-\mathbf{r}'|}{a}\right)e^{\frac{|\mathbf{r}-\mathbf{r}'|}{a}}\, d^3r\, d^3r', \tag{5.77}
$$

while B_3 corresponds to the relative Coulomb energy of an uniformly charged, deformed body

$$
B_3 = \frac{15}{32\pi^2}\frac{A^{-5/3}}{r_0^5}\int_V\int_{V'}\frac{d^3r\, d^3r'}{|\mathbf{r}-\mathbf{r}'|}\left[1 - (1 + \frac{|\mathbf{r}-\mathbf{r}'|}{2a_{\mathrm{den}}})e^{-|\mathbf{r}-\mathbf{r}'|a_{\mathrm{den}}}\right]. \tag{5.78}
$$

The following parameters of the Yukawa plus exponential model were adjusted to the experimental masses of nuclei and the fission barrier heights:

$a_{\mathrm{den}} = 0.99/2^{1/2}$ fm	range of the charge distribution function,
$a_{\mathrm{el}} = 1{,}433 \cdot 10^{-5}$ MeV	electron binding constant,
$\Delta = 12$ MeV	pairing energy gap constant,
$\delta = 20$ MeV	pairing asymmetry constant,
$r_{\mathrm{p}} = 0.80$ fm	proton radius,
$r_0 = 1.16$ fm	nucleus radius constant,
$a = 0.68$ fm	Yukawa potential range,
$a_{\mathrm{s}} = 21.13$ MeV	surface energy constant,
$\kappa_{\mathrm{s}} = 2.3$	surface asymmetry constant,
$a_{\mathrm{v}} = 15.9937$ MeV	volume energy constant,
$\kappa_{\mathrm{v}} = 1.927$	volume symmetry constant,
$W = 36$ MeV	Wigner constant,
$c_0 = 4.4$ MeV	constant at A^0,
$c_a = 0.212$ MeV	charge asymmetry constant.

The following section shows another, much simpler modern LD model that reproduces well the macroscopic properties of nuclei known at present.

5.10 LUBLIN-STRASBOURG LIQUID DROP MODEL

After the development of more and more complicated formulas for the macroscopic part of nuclear energy, it appeared that it is enough to add the curvature term to the liquid drop expression of Myeres and Swiatecki (5.55) in order to reproduce well both nuclear masses and fission barrier heights. The Lublin-Strasbourg Drop (LSD) formula developed in 2011 reads:

$$
\begin{aligned}
E_{\mathrm{LSD}} &= ZM_H + NM_n - 0.00001433Z^{2.39} \\
&-b_{\mathrm{vol}}(1-\kappa_{\mathrm{vol}}I^2)A \\
&+b_{\mathrm{surf}}(1-\kappa_{\mathrm{surf}}I^2)A^{2/3} \\
&+b_{\mathrm{curv}}(1-\kappa_{\mathrm{curv}}I^2)A^{1/3} \\
&+\frac{3}{5}e^2\frac{Z^2}{r_0^{\mathrm{ch}}A^{1/3}} - C_4\frac{Z^2}{A} - 10\cdot\exp(-42|I|/10)\,,
\end{aligned}
\tag{5.79}
$$

where $I = (N - Z)/A$. The last term is the congruence (Wigner type) energy taken from the estimates of Myers and Swiatecki.
The parameters of the LSD formula are following:

$$\begin{aligned} b_{\rm vol} &= 15.4920\ \text{MeV}\,, & \kappa_{\rm vol} &= 1.8601\,, \\ b_{\rm surf} &= 16.9707\ \text{MeV}\,, & \kappa_{\rm surf} &= 2.2938\,, \\ b_{\rm curv} &= 3.8602\ \text{MeV}\,, & \kappa_{\rm curv} &= -2.3764\,, \\ r_0^{\rm ch} &= 1.21725\ \text{fm}\,, & C_4 &= 0.91810\ \text{MeV}\,. \end{aligned} \tag{5.80}$$

They were fitted to the experimental masses of nuclei known in 2003. After including the microscopic (shell and pairing) energy corrections, this formula reproduces all available in the 2020-year nuclear masses with r.m.s. deviation equal 0.69 MeV. It also predicts the fission barrier heights close to their experimental values for 80 nuclei. One has to stress here that none of the LSD parameters was fitted to the barrier heights. The simplicity and a relatively small number of adjustable LSD parameters gives hope for reliable prediction of properties of not yet discovered nuclei.

CHAPTER 6

Average single-particle potentials

Constant binding energy per nucleon in the nucleus indicates nuclear forces' saturation and their short-range nature. The short range of nuclear forces causes nucleons inside the nucleus to be equally attracted. On the other hand, near the nuclear surface, the nucleons are attracted less due to the smaller number of neighbors. This effect can be described by introducing the average single-particle potential felt by a single particle and created by the rest of the nucleons. The shell model is based on this assumption. Eigenstates of Hamiltonian with average potential should group in shells corresponding to the magic numbers (confer Sec. 5.4). Replacement of two-body nuclear interactions with other surrounded nucleons by average single-particle potential is justified by the 'healing' Bethe-Goldstone wave function (Fig. 3.1), which appears due to the Pauli exclusion principle.

The average potential depth (V_0) has already been estimated in Sec. 4.2.1, knowing the nucleon separation energy ($S_{\mathrm{n(p)}}$) and the Fermi energy (e_F) calculated in the Fermi gas model. If $S_{\mathrm{n(p)}} \approx 8$ MeV and $e_\mathrm{F} \approx 39$ MeV, then $V_0 \approx 47$ MeV. The potential should also be correlated with the distribution density, i.e., inside the nucleus, it should be almost constant and gradually disappear at the surface. Fig. 6.1 shows the most commonly used types of single-particle potentials in nuclei: the rectangular well, the Woods-Saxon potential, and the harmonic oscillator potential.

The harmonic oscillator's potential has many important applications in nuclear physics. Its eigenfunctions are frequently used as the base in which Hamiltonians with more realistic (e.g., Nilsson or Woods-Saxon) potentials are diagonalized.

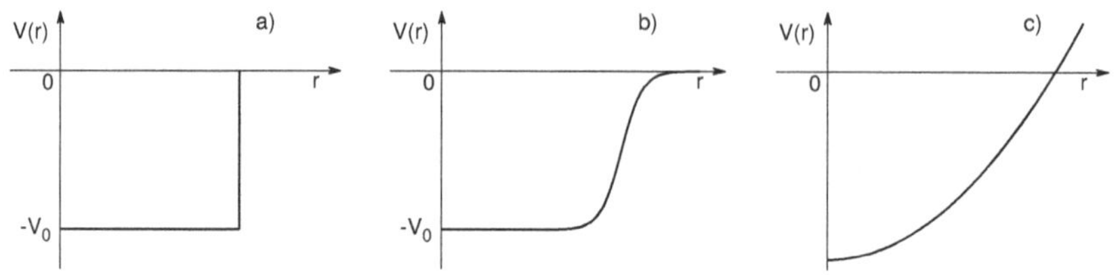

Figure 6.1 Average single-particle potentials: a) square well, b) Woods-Saxon, c) harmonic oscillator.

6.1 HARMONIC OSCILLATOR POTENTIAL

Let us solve the Schroedinger equation for Hamiltonian with kinetic energy and harmonic oscillator potential

$$\hat{H} = \frac{1}{2M}\hat{p}^2 + \frac{1}{2}M\mathring{\omega}_0^2 q^2 , \tag{6.1}$$

where $q = \{q_1, q_2, q_3, \ldots, q_n\}$ (e.g. $\{x, y, z\}$) represents the coordinates in n-dimensional space.

In the following, the formalism of the occupation number representation will be used. Let us define the boson creation $\hat{b}_k^+$ and annihilation $\hat{b}_k$ operators of oscillations quanta (phonons):

$$\begin{aligned} \hat{b}_k^+ &\equiv \frac{1}{\sqrt{2M\hbar\mathring{\omega}_0}}(\hat{\mathrm{p}}_k + iM\mathring{\omega}_0 q_k) , \\ \hat{b}_k &\equiv \frac{1}{\sqrt{2M\hbar\mathring{\omega}_0}}(\hat{\mathrm{p}}_k - iM\mathring{\omega}_0 q_k) , \end{aligned} \tag{6.2}$$

where M is the mass of a particle and $\mathring{\omega}_0$ the oscillation frequency of the spherical oscillator. Of course, the operators $\hat{b}_k^+$ and $\hat{b}_k$ meet the boson commutation rules:

$$[\hat{b}_i, \hat{b}_j^+] = \hat{b}_i\hat{b}_j^+ - \hat{b}_j^+, \hat{b}_i = \delta_{ij} \quad \text{and} \quad [\hat{b}_i, \hat{b}_j] = [\hat{b}_i^+, \hat{b}_j^+] = 0 . \tag{6.3}$$

The vacuum state $|0\rangle$, i.e., state without phonons, is defined as follows:

$$\hat{b}_k|0\rangle = 0 \quad \text{and} \quad \langle 0|0\rangle = 1 . \tag{6.4}$$

One boson state $|1\rangle$ is determined by:

$$\hat{b}_k^+|0\rangle = |1\rangle . \tag{6.5}$$

It is normalized:

$$\langle 0|\hat{b}_k\hat{b}_k^+|0\rangle = \langle 0|1+\hat{b}_k^+\hat{b}_k|0\rangle = \langle 0|0\rangle = 1\,. \tag{6.6}$$

The two boson states $\hat{b}_k^+\hat{b}_k^+|0\rangle$ have the norm 2, which is easy to prove using the commutation rules (6.3) and properties of the vacuum state

$$\langle 0|\hat{b}_k\hat{b}_k\hat{b}_k^+\,\hat{b}_k^+|0\rangle = 2\langle 0|\hat{b}_k\hat{b}_k^+|0\rangle = 2\,. \tag{6.7}$$

Let us induce the normalizing factor $\mathcal{N}(n)$ of n phonon state $|n\rangle$:

$$\mathcal{N}(1) = 1, \quad \mathcal{N}(2) = 2\mathcal{N}(1)\,.$$

It is easy to write a general relationship using the commutation rules of boson operators,

$$\mathcal{N}(n) = n\mathcal{N}(n-1) = n(n-1)\mathcal{N}(n-2) = \ldots,$$

hence

$$\mathcal{N}(n) = n!\,.$$

So, the normalized n bosons state reads:

$$|n\rangle = \frac{1}{\sqrt{n!}}(\hat{b}_k^+)^{\mathrm{n}}|0\rangle\,. \tag{6.8}$$

Using the above formula, one can show that the operator $\hat{b}_k^+$ in action on the state $|n\rangle$ gives

$$\hat{b}_k^+|n\rangle = \sqrt{n+1}\,|n+1\rangle\,, \tag{6.9}$$

because

$$\hat{b}_k^+\frac{(\hat{b}_k^+)^n}{\sqrt{n!}}|0\rangle = \sqrt{n+1}\frac{(\hat{b}_k^+)^{n+1}}{\sqrt{(n+1)!}}|0\rangle\,.$$

Similarly, one has

$$\hat{b}_k|n\rangle = \sqrt{\mathrm{n}}\,|n-1\rangle \tag{6.10}$$

because

$$\hat{b}_k\frac{(\hat{b}_k^+)^n|0\rangle}{\sqrt{n!}} = \sqrt{\mathrm{n}}\frac{(\hat{b}_k^+)^{n-1}}{\sqrt{(n-1)!}}|0\rangle\,.$$

The operator $\hat{b}_k^+\hat{b}_k$ does not change the number of phonons, and its expected value is equal to the number of k bosons in the state

$$\langle n|\hat{b}_k^+\hat{b}_k|n\rangle = n\,,$$

which can be easily demonstrated using formulas (6.9) and (6.10). The total phonon number operator is defined as a sum

$$\hat{\mathrm{n}} = \sum_{k=1}^{3} \hat{b}_k^+ \hat{b}_k . \tag{6.11}$$

The harmonic oscillator Hamiltonian (6.1) can now be expressed in the form

$$\hat{H} = \hbar \mathring{\omega}_0 \sum_{k=1}^{3} \left(\hat{b}_k^+ \hat{b}_k + \frac{1}{2} \right) . \tag{6.12}$$

Its eigenfunction can be presented as a product

$$\psi_{n_1 n_2 n_3} = X_{n_1}(x) Y_{n_2}(y) Z_{n_3}(z) , \tag{6.13}$$

where X, Y, Z functions are Hermite polynomials dependent on x, y, z respectively. Quantum numbers describing the state of $\psi_{n_1 n_2 n_3} \equiv |n_1 n_2 n_3\rangle$, $n_i = 0, 1, 2, \ldots$, correspond to the number of phonons in each of the three directions.

Introducing the total number of oscillation quanta $N = n_1 + n_2 + n_3$, one can write

$$\hat{H} |n_1 n_2 n_3\rangle = \left(N + \frac{3}{2} \right) \hbar \mathring{\omega}_0 |n_1 n_2 n_3\rangle . \tag{6.14}$$

So, one has got a spectrum of discrete (i.e., quantized) eigenenergies of $\hat{H}$

$$e_{\mathrm{n}} = \left(N + \frac{3}{2} \right) \hbar \mathring{\omega}_0 \tag{6.15}$$

with the following level of degeneration:

$$\deg(N) = \sum_{n_3=0}^{\mathrm{N}} \sum_{n_2=0}^{N-n_3} 1 = \sum_{n_3=0}^{\mathrm{n}} (N - n_3 + 1) = \frac{(N+1)(N+2)}{2} . \tag{6.16}$$

Due to two possible spin orientations, it can be found at max $2 \cdot \deg(N)$ nucleons of one type on each shell N. This number, as well as the total number $\mathcal{L}$ of nucleons on all shells up to the given N, are shown in Table 6.1.

So, the spectrum of equidistant degenerated levels forming the shells as shown in Fig. 6.2 is received. The magic numbers corresponding to the closed (i.e., filled by nucleons) shells of the oscillator's potential are as follows:

$$\mathcal{L}(N) = 2 \sum_{N'=0}^{\mathrm{N}} \frac{(N'+1)(N'+2)}{2} = \frac{1}{3}(N+1)(N+2)(N+3) \simeq \frac{(N+2)^3}{3} . \tag{6.17}$$

Table 6.1 Nucleon distribution on the harmonic oscillator potential shells.

N	0	1	2	3	4	5
$2 \cdot \deg(N)$	2	6	12	20	30	42
$\mathcal{L}(N)$	2	8	20	40	70	112

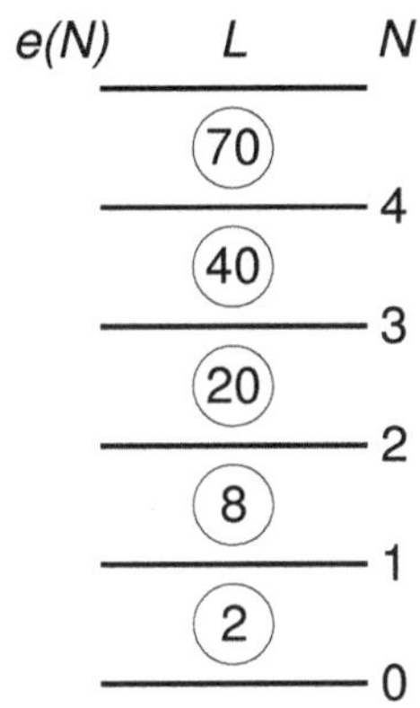

Figure 6.2 Spectrum of eigenenergies of the spherical harmonic oscillator Hamiltonian.

The harmonic oscillator potential reproduces only the first three experimental (see Sec. 5.4) magic numbers 2, 8, and 20.

Eq. (6.17) allows to obtain an approximate relationship between the mass number A and the main quantum number N of the highest occupied shell. For $Z = N = A/2$ the highest occupied number shell N_{max} is associated with A

$$A = \mathcal{L}(N_{\text{max}}) \simeq \frac{2}{3}(N_{\text{max}} + 2)^3$$

or otherwise

$$N_{\text{max}} \simeq \left(\frac{3}{2}A\right)^{1/3} - 2. \tag{6.18}$$

Let us now estimate the only parameter of the harmonic oscillator potential, i.e., its frequency $\overset{\circ}{\omega}_0$. For this purpose, one has to evaluate the mean value of the square radius in each state $|n_1 n_2 n_3\rangle$ and then compare its average with the experimental value of nuclear square radius (4.7). Using the definition (6.2), one can express the coordinate q_k by the creation and annihilation operators of bosons

$$q_k = -i\sqrt{\frac{\hbar}{2M\overset{\circ}{\omega}_0}}(\hat{b}_k^+ - \hat{b}_k) \tag{6.19}$$

and

$$q_k^2 = \frac{\hbar}{2M\overset{\circ}{\omega}_0}(-\hat{b}_k^+\hat{b}_k^+ + 2\hat{b}_k^+\hat{b}_k + 1 - \hat{b}_k\hat{b}_k), \tag{6.20}$$

hence

$$\langle n_k|q_k^2|n_k'\rangle = \frac{\hbar}{2M\overset{\circ}{\omega}_0}\begin{cases} 2n_k + 1 & \text{for } n_k = n_k', \\ -\sqrt{n_k(n_k-1)} & \text{for } n_k = n_k' + 2, \\ -\sqrt{(n_k+1)(n_k+2)} & \text{for } n_k = n_k' - 2. \end{cases} \tag{6.21}$$

Diagonal matrix element of operator

$$r^2 = \sum_{k=1}^{3} q_k^2 \tag{6.22}$$

is therefore equal

$$\langle n_1n_2n_3|r^2|n_1n_2n_3\rangle = \frac{\hbar}{M\overset{\circ}{\omega}_0}(N + \frac{3}{2}). \tag{6.23}$$

The expected value of the square radius of the nucleus with the mass number A is therefore

$$\begin{aligned}\sum_{i=1}^{A}\langle r_i^2\rangle &= \frac{\hbar}{M\overset{\circ}{\omega}_0}\sum_{i=1}^{A}(N_i + \frac{3}{2}) \\ &= \sum_{N=0}^{N_{\max}} 2\cdot 2(N + \frac{3}{2})\frac{(N+1)(N+2)}{2}\frac{\hbar}{M\overset{\circ}{\omega}_0} \\ &\simeq \frac{1}{2}(N_{\max}+2)^4 \cdot \frac{\hbar}{M\overset{\circ}{\omega}_0}.\end{aligned} \tag{6.24}$$

Equating the above formula with the similar expression obtained for a homogeneous mass distribution

$$\sum_{i=1}^{A}\langle r_i^2\rangle = \frac{3}{5}AR^2 = \frac{3}{5}r_0^2A^{5/3} \tag{6.25}$$

allows to estimate the $\overset{\circ}{\omega}_0$ parameter. Since the relationship (6.18) shows that $N_{\max} + 2 \approx (\frac{3}{2}A)^{1/3}$, we'll get the equation

$$\frac{\hbar}{2M\overset{\circ}{\omega}_0}\left(\frac{3}{2}A\right)^{4/3} = \frac{3}{5}r_0^2 A^{5/3}, \tag{6.26}$$

which for $r_0 = 1.2$ fm (4.7) gives

$$\hbar\overset{\circ}{\omega}_0 = \frac{(\hbar c)^2}{Mc^2 r_0^2} \cdot \frac{3^{1/3} \cdot 5}{2^{1/3} \cdot 4} A^{-1/3} \approx \frac{41.6}{A^{1/3}}\mathrm{MeV}. \tag{6.27}$$

Exact matching $\langle r^2 \rangle$ to known radii of nuclei, also including those with not equal number of protons and neutrons $Z \neq N \neq A/2$ gives:

$$\hbar\overset{\circ}{\omega}_0 = 40/A^{1/3}\ \mathrm{MeV}. \tag{6.28}$$

Unfortunately, as one has already shown, the harmonic oscillator's potential cannot reproduce all known magic numbers and needs improvement. One can deepen the oscillator's potential at the edge of the nucleus and consider the coupling of orbital and spin motion. It is convenient to make it in the spherical coordinate system. Because the Hamiltonian oscillator $\hat{H}$ (6.1) has spherical symmetry, its eigenequation can be split into the radial and angular parts when one takes its eigenfunctions in the product form

$$\psi_{Nlm_l} = \frac{R_{Nl}(r)}{r} Y_{lm_l}(\vartheta, \varphi), \tag{6.29}$$

where

$$N = 0, 1, 2, \ldots, \quad l = N, N-2, N-4, \ldots 0(1), \quad m_l = -l, (-l+1), \ldots l.$$

The functions $Y_{lm_l}(\varphi, \vartheta)$ are spherical harmonics (2.10), and the radial functions $R_{Nl}(r)$ satisfy the equation:

$$-\frac{\hbar^2}{2M}\frac{\partial^2}{\partial r^2} R_{Nl} + \left(\frac{1}{2}M\,\overset{\circ}{\omega}{}_0^2\, r^2 - E + \frac{\hbar^2 l(l+1)}{2Mr^2}\right) R_{Nl} = 0, \tag{6.30}$$

which leads to the following solution:

$$R_{Nl}(r) = \frac{\beta^{3/2}\Gamma\left(\frac{N+l+3}{2}\right) + 2}{\left[\Gamma\left(l + \frac{3}{2}\right)\right]^2 \left(\frac{N-l}{2}\right)!} (\beta r^2)^{1/2} \mathrm{e}^{-\frac{\beta r^2}{2}} \cdot F\left(-\frac{N-l}{2}, l + \frac{3}{2}, \beta r^2\right), \tag{6.31}$$

where $\beta \equiv M \overset{\circ}{\omega}_0 / \hbar$ and

$$F(a,c,x) = 1 + \frac{a}{c}x + \frac{1}{2!}\frac{a(a+1)}{c(c+1)}x^2 + \frac{1}{3!}\frac{a(a+1)(a+2)}{c(c+1)(c+2)}x^3 + \ldots$$

is the confluent hyper-geometric function.
The Γ Euler functions are:

$$\Gamma(x) = \int_0^{\infty} e^{-t} t^{x-1} dt. \tag{6.32}$$

The Hamiltonian with harmonic oscillator has been made more realistic by Goeppert-Mayer and Jensen, and later by Nilsson, who have supplemented it with the terms $\hat{\mathbf{l}} \cdot \hat{\mathbf{s}}$ describing the coupling of orbital and spin motion, and the $\hat{\mathbf{l}}^2$ term to make the potential deeper for nucleons with high orbital angular momenta

$$\hat{H}_{\text{sp}} = -\frac{\hbar^2}{2M}\hat{\Delta} + \frac{1}{2}M\overset{\circ}{\omega}{}_0^2 r^2 + V_{ls} \cdot (\hat{\mathbf{l}} \cdot \hat{\mathbf{s}}) + V_{l^2} \cdot [\hat{\mathbf{l}}^2 - \langle \hat{\mathbf{l}}^2 \rangle_N] \,. \tag{6.33}$$

One can solve the eigenproblem of the above single-particle Hamiltonian by taking its eigenfunctions as a product of the eigenfunctions ψ_{Nlm_l} (6.29) of Hamiltonian with harmonic oscillator $\hat{H}$ (6.12) and the spin functions χ_{sm_s}. In this base of functions, the term $\hat{\mathbf{l}} \cdot \hat{\mathbf{s}}$ has the following matrix elements:

$$\begin{aligned} \langle N,l,m_l \pm 1, \mp\frac{1}{2} | \, \hat{\mathbf{l}} \cdot \hat{\mathbf{s}} \, | N',l',m_l, \pm\frac{1}{2} \rangle &= \delta_{NN'}\,\delta_{ll'}\,\frac{1}{2}\,\sqrt{(l \mp m_l)(l \pm m_l + 1)}, \\ \langle N,l,m_l, \pm\frac{1}{2} | \, \hat{\mathbf{l}} \cdot \hat{\mathbf{s}} \, | N',l',m_l, \pm\frac{1}{2} \rangle &= \pm\frac{1}{2} m_l \,\delta_{NN'}\,\delta_{ll'} \,. \end{aligned} \tag{6.34}$$

The functions with the well-defined total angular momentum $\hat{\mathbf{j}} = \hat{\mathbf{l}} + \hat{\mathbf{s}}$ will be the eigenfunctions of $\hat{H}_{\text{sp}}$ because

$$\hat{\mathbf{l}} \cdot \hat{\mathbf{s}} = \frac{1}{2}(\hat{\mathbf{j}}^2 - \hat{\mathbf{l}}^2 - \hat{\mathbf{s}}^2). \tag{6.35}$$

Such functions can be built with eigenfunctions of the spherical oscillator ψ_{Nlm_l} and spin functions χ_{sm_s}:

$$|Njlsm\rangle = \sum_{m_l + m_s = m} (lm_l s m_s | jm) \psi_{Nlm_l} \chi_{sm_s} \,, \tag{6.36}$$

where $(lm_l sm_s|jm)$ is the Clebsch-Gordan coefficient. In this base, the eigenequation $\hat{H}_{sp}$ has the form

$$\hat{H}_{sp}|Njlsm\rangle = \Big\{(N+3/2)\hbar\overset{\circ}{\omega}_0 + V_{ls}[j(j+1) - l(l+1) - s(s+1)]/2 + V_{l^2}[l(l+1) - \langle\hat{\mathbf{l}}^2\rangle_N]\Big\}|Njlsm\rangle. \tag{6.37}$$

Noticing that $s = \frac{1}{2}$ and $j = l \pm \frac{1}{2}$ one can evaluate the expected value of the spin-orbit term

$$\frac{1}{2}\left[j(j+1) - l(l+1) - \frac{3}{4}\right] = \begin{cases} l/2 & \text{for } j = l + \frac{1}{2} \\ -(l+1)/2 & \text{for } j = l - \frac{1}{2}. \end{cases} \tag{6.38}$$

Introducing the standard Nilsson notation:

$$V_{ls} \equiv -2\kappa\hbar\overset{\circ}{\omega}_0, \qquad V_{l^2} \equiv -\kappa\mu\hbar\overset{\circ}{\omega}_0 \tag{6.39}$$

we obtain the final form of the Nilsson Hamiltonian eigenequation:

$$\hat{H}_{sp}|Njlsm\rangle = \Big\{\hbar\overset{\circ}{\omega}_0 (N + \frac{3}{2}) + +\kappa\hbar\overset{\circ}{\omega}_0\left[\begin{pmatrix} l+1 & \text{for } j = l - \frac{1}{2} \\ -l & \text{for } j = l + \frac{1}{2} \end{pmatrix} - \mu l(l+1)\right]\Big\}|Njlsm\rangle. \tag{6.40}$$

The μ and κ parameters are matched to the experimental distances between single-particle levels for protons (p) and neutrons (n) and are, for example, equal to: $\kappa_p = \kappa_n = 0.0637$, $\mu_p = 0.637$, $\mu_n = 0.42$ for nuclei with $A \approx 160$. The system of single-particle levels obtained with such parameters well reproduces all known magic numbers.

6.2 DEFORMED NILSSON POTENTIAL

6.2.1 Average potential shape parameterization

In microscopic calculations for non-spherical nuclei, the shape of the single-particle potential wells must correspond to the spatial nucleon density distribution in a nucleus. Therefore, spheroidally deformed nuclei are often described by the potential of an anisotropic harmonic oscillator

$$V = \frac{1}{2}M\left[\omega_x^2 x^2 + \omega_y^2 y^2 + \omega_z^2 z^2\right]. \tag{6.41}$$

When the axial symmetry of the nucleus is maintained, the oscillator frequencies along its main axes are equal $\omega_x = \omega_y \equiv \omega_\perp$. Nilsson used at first the so-called δ parameterization to describe quadrupole deformation

$$
\begin{aligned}
\omega_\perp^2 &= \omega_0^2(\delta)[1 + \frac{2}{3}\delta], \\
\omega_z^2 &= \omega_0^2(\delta)[1 - \frac{4}{3}\delta],
\end{aligned}
\tag{6.42}
$$

where ω_0 is the frequency of the deformed oscillator and δ describes the elongation of the axially symmetric spheroid corresponding to the equipotential surface. Note that all equipotential surfaces are mathematically similar in the anisotropic harmonic oscillator potential. However, another Nilsson parameterization, the ε-one, turned out to be more practical

$$
\begin{aligned}
\omega_\perp &= \omega_0(\varepsilon)\left[1 + \frac{1}{3}\varepsilon\right], \\
\omega_z &= \omega_0(\varepsilon)\left[1 - \frac{2}{3}\varepsilon\right].
\end{aligned}
\tag{6.43}
$$

Namely, it allows for a relatively simple determination in an analytical way, i.e., without matrix diagonalization, of eigenfunctions and eigenvalues of the anisotropic harmonic oscillator. Later, in the process of extending the scope of applicability of Nilsson's potential for more complex shapes of nuclei, the quadrupole deformation ε was supplemented by deformation parameters of other multipolarities (confer Sec. 6.2.3):

$$
\varepsilon_\lambda \equiv \varepsilon_1, \varepsilon, \varepsilon_3, \varepsilon_4, \varepsilon_5, \dots .
$$

In the case of axial symmetry of the nucleus, parameters ε_λ with λ even describe shapes symmetric with respect to the $x0y$ plane, while for λ odd, the shapes breaking this mirror symmetry. Typical shapes corresponding to the different multipolarity deformations are shown in Fig. 6.3. In general, the number of 'horns' in the surface cross-section is equal to the deformation multipolarity. The ε_1 parameter moves the center of gravity of the nucleus along the z axis, and it is set in such a way that for $\varepsilon_\lambda \neq 0$ the mass center of the nucleus is located at the beginning of the coordinate system.

Among nuclei found in nature (Fig. 5.2), except spherical ones, the quadrupole ($\varepsilon \neq 0$) and hexadecapole deformed ($\varepsilon_4 \neq 0$) nuclei are the most common. Pear-like nuclei with non-zero octupole deformation ($\varepsilon_3 \neq 0$) appear less frequently. The same is true for those corresponding to the non-axial or more sophisticated shapes.

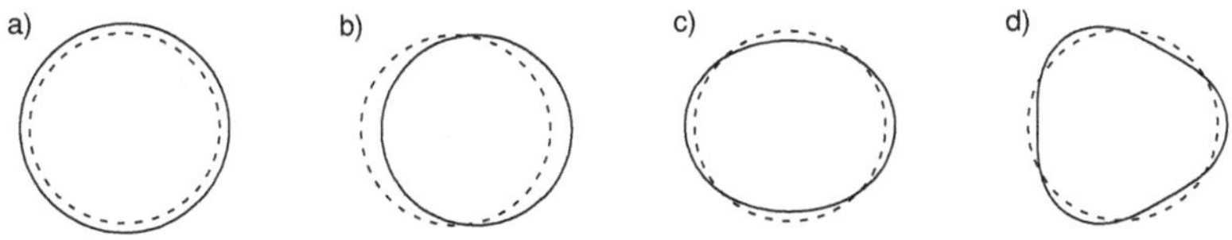

Figure 6.3 Schematic shapes of axially symmetric nuclei corresponding to deformation of different multipolarity: a) monopole - $\lambda = 0$, b) dipole - $\lambda = 1$, c) quadrupole - $\lambda = 2$, d) octupole - $\lambda = 3$.

If a nucleus is not axially symmetric with respect to z-axis, an additional parameters should be introduced. In the case of quadrupole deformation, the main axes of the spheroidal nucleus can then be parameterized as follows

$$\begin{aligned}
\omega_x &= \omega_0(\varepsilon,\gamma)\left[1+\frac{2}{3}\varepsilon\cos\left(\gamma-\frac{\pi}{3}\right)\right], \\
\omega_y &= \omega_0(\varepsilon,\gamma)\left[1+\frac{2}{3}\varepsilon\cos\left(\gamma+\frac{\pi}{3}\right)\right], \\
\omega_z &= \omega_0(\varepsilon,\gamma)\left[1-\frac{2}{3}\varepsilon\cos\gamma\right].
\end{aligned} \tag{6.44}$$

For $\gamma = 0$, this parametrization is the same as in Eq. (6.43). It should be emphasized that one follows nuclear shapes corresponding to these types of deformation only by choosing a specific parameterization. It is reasonable to choose such a deformation parameter set that makes the calculation simple. It depends, of course, on a problem one would like to discuss. The above choice enables a particular transformation of coordinates (see next section), which brings the matrix of the single-particle Hamiltonian into a quasi-diagonal form. When determining the frequency ω_0, one uses the volume conservation condition due to the incompressibility of nuclear matter. This condition demands that the volume of the spherical nucleus be the same as deformed. In the Nilsson Hamiltonian, it can be only approximately achieved assuming that the volume inside each equipotential surface is constant, i.e., does not change with deformation.

6.2.2 Diagonalization of the Nilsson Hamiltonian

The Nilsson Hamiltonian consists of the kinetic energy operator, the anisotropic harmonic oscillator potential, symmetric with respect to rotation around the z-axis (V_{osc}), and the corrective term (V_{corr}) describing: spin-orbit interaction $\hat{\mathbf{l}}\cdot\hat{\mathbf{s}}$ and one proportional to $\hat{\mathbf{l}}^2$, flattening the bottom of the potential well. From $\hat{\mathbf{l}}^2$ is subtracted the mean value of l^2 in a given shell ($\langle\hat{\mathbf{l}}^2\rangle_N$) to keep on the average equal distances between the 'centers of gravity' of subsequent shells. The presence of Coulomb potential for protons was included here by an appropriate change of the harmonic oscillator stiffness parameter concerning that for neutrons.

The Nilsson Hamiltonian's eigenproblem

$$\hat{H}_{\rm sp} = -\frac{\hbar^2\hat{\Delta}}{2M} + \frac{M}{2}\left[\omega_\perp^2(x^2+y^2)+\omega_z^2z^2\right] + V_{ls}\cdot\hat{\mathbf{l}}\cdot\hat{\mathbf{s}} + V_{l^2}\left[\hat{\mathbf{l}}^2 - \langle\hat{\mathbf{l}}^2\rangle_N\right]. \quad (6.45)$$

is solved by the diagonalization of its matrix representation based on eigenfunctions of the spherical harmonic oscillator Hamiltonian (6.1). These functions may be labeled with the following Nilssons' quantum numbers (6.29)

$$|N,l,\Lambda,\Sigma\rangle \equiv |i\rangle,$$

where

$$\begin{aligned} &N = 0,1,2,\ldots,N_{\rm max}, \\ &l = N, N-2, N-4,\ldots,0(1), \\ &\Lambda \equiv m_l = -l,-l+1,\ldots,l, \\ &\Sigma \equiv m_s = \pm\tfrac{1}{2}. \end{aligned} \quad (6.46)$$

The next step is to bring the matrix corresponding to $\mathrm{H}_{\rm sp}$

$$(\mathrm{H}_{\rm sp})_{ij} = \langle i|\hat{H}_{\rm sp}|j\rangle \quad (6.47)$$

into quasi-diagonal form.

The mathematical procedures by Nilsson rely on an appropriate selection of coordinates, which allows to express the matrix elements (6.47) in a particularly simple way. Using the ε parameterization (6.43) for the anisotropic oscillator frequency and κ,μ (6.39) parameters for the correction term, one gets $\hat{H}_{\rm sp}$ (6.33) in the form

$$\begin{aligned} \hat{H}_{\rm sp} &= \frac{-\hbar^2\hat{\Delta}}{2M} + \frac{M\omega_0^2(\varepsilon)}{2}\left[(1+\frac{\varepsilon}{2})^2(x^2+y^2)+(1-\frac{2}{3}\varepsilon)^2z^2\right] \\ &-\hbar\mathring{\omega}_0\kappa\left\{2\hat{\mathbf{l}}\cdot\hat{\mathbf{s}}+\mu\left[l^2+\langle l^2\rangle_N\right]\right\}, \end{aligned} \quad (6.48)$$

where $\mathring{\omega}_0$ is the oscillation frequency of the spherical harmonic oscillator

$$\mathring{\omega}_0 \equiv \omega_0(0). \quad (6.49)$$

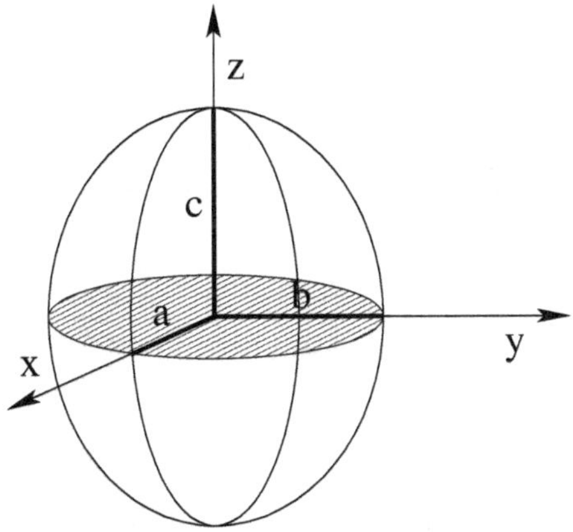

Figure 6.4 Spheroidal deformation of a nucleus.

The oscillation frequency $\omega_0(\varepsilon)$ is determined from the volume conservation condition inside the equipotential surface V_{osc} = const. This condition provides approximately a constant volume of the deformed nucleus.

In the spheroidal nucleus, the half-axes are defined by formulas

$$a = b = R_0 \frac{\omega_0}{\omega_\perp}, \quad c = R_0 \frac{\omega_0}{\omega_z}, \tag{6.50}$$

where R_0 is the radius of the spherical nucleus.
So, the following condition is sufficient to preserve the volume:

$$\frac{4}{3}\pi R_0^3 = \frac{4}{3}\pi a^2 c \tag{6.51}$$

which leads to the relationship between the anisotropic oscillator frequencies

$$\overset{\circ}{\omega}{}_0^3 = \omega_\perp^2 \omega_z, \tag{6.52}$$

or

$$\frac{\overset{\circ}{\omega}_0}{\omega_0(\varepsilon)} = \left(1 + \frac{1}{3}\varepsilon\right)^{2/3} \left(1 - \frac{2}{3}\varepsilon\right)^{1/3}. \tag{6.53}$$

Fig. 6.4 schematically shows the shape of deformed nuclei.

The mean value of the angular momentum square $\langle \hat{\mathbf{l}}^2 \rangle_N$ occurring in (6.48), is the sum of $l(l+1)$ over all single-particle states within one shell

N divided by the number of the states in the shell:

$$\langle \hat{\mathbf{l}}^2 \rangle_N = \frac{\sum\limits_{l=N,N-2...}^{0(1)} \sum\limits_{\Lambda=-l,-l+1...}^{l} l(l+1)}{\sum\limits_{l=N,N-2...}^{0(1)} \sum\limits_{\Lambda=-l,-l+1...}^{l} 1} = \frac{N(N+3)}{2}. \tag{6.54}$$

To diagonalize the Hamiltonian (6.48) Nilsson has introduced dimensionless stretched coordinates

$$x = \sqrt{\frac{\hbar}{M\omega_\perp}}\xi, \qquad y = \sqrt{\frac{\hbar}{M\omega_\perp}}\eta, \qquad z = \sqrt{\frac{\hbar}{M\omega_z}}\zeta. \tag{6.55}$$

After this transformation, when one omits terms higher than linear in ε, the Nilsson Hamiltonian takes the following form:

$$\begin{aligned} \hat{H}_{\mathrm{sp}} &= \frac{1}{2}\hbar\omega_0(\varepsilon)\left[-\hat{\Delta}_t + \rho^2\right] \\ &+\frac{1}{6}\hbar\omega_0(\varepsilon)\,\varepsilon\left[\left(-\frac{\partial^2}{\partial\xi^2}+\xi^2\right)+\left(-\frac{\partial^2}{\partial\eta^2}+\eta^2\right)-2\left(-\frac{\partial^2}{\partial\zeta^2}+\zeta^2\right)\right] \\ &-\hbar\overset{\circ}{\omega}_0\,\kappa\left[\hat{\mathbf{l}}_t\cdot\hat{\mathbf{s}}+\mu\left(\hat{\mathbf{l}}_t^2-\frac{N(N+3)}{2}\right)\right], \end{aligned} \tag{6.56}$$

where $\rho^2 = \xi^2+\eta^2+\zeta^2$, and the subscript t means that the operators are specified in the stretched coordinates.

The terms in the second line of the above equation can be grouped into two parts:

$$\xi^2+\eta^2-2\zeta^2 \equiv -2\rho^2 P_2(\cos\vartheta_t) \tag{6.57}$$

and

$$2\frac{\partial^2}{\partial\zeta^2}-\frac{\partial^2}{\partial\xi^2}-\frac{\partial^2}{\partial\eta^2} \equiv 2p^2 P_2(\cos\vartheta_{tp}), \tag{6.58}$$

where P_2 is the Legendre polynomial, p is the momentum operator, and ϑ_{tp} the angle between the vector $\mathbf{p}$ and the axis p_ζ. Diagonal matrix elements of (6.57) and (6.58) between the states from the same N shell are equal, while the non-diagonal elements have opposite signs and their sum reduces to 0. So, the following simplified operator:

$$\hat{H}_{\mathrm{sp}} = \frac{1}{2}\hbar\omega_0(\varepsilon)\left[-\hat{\Delta}_t+\rho^2\right]-\frac{\varepsilon}{3}\hbar\omega_0(\varepsilon)\rho^2 P_2(\cos\vartheta_t)\delta_{NN'}+V^{\mathrm{t}}_{\mathrm{corr}}, \tag{6.59}$$

has the same matrix elements as the Nilsson Hamiltonian (6.56). The term $V^{\mathrm{t}}_{\mathrm{corr}}$ is the correction term in the stretched coordinates. The obtained Hamiltonian has non-vanishing matrix elements between the states from the same N-shell only. It also commutes with the third component of angular momentum,

$$[\hat{H}_{\mathrm{sp}}, \hat{j}_z] = 0\,, \tag{6.60}$$

so $\Omega = \Lambda + \Sigma$ is a good quantum number.

To evaluate the matrix elements of $\hat{H}_{\mathrm{sp}}$ in the basis (6.46) it is sufficient to use Eq. (6.34) and obtain the following formulas:

$$\begin{aligned}
&\langle N,l,\Lambda,\Sigma|\rho^2|N,l,\Lambda,\Sigma\rangle = \langle N,l,\Lambda,\Sigma| - \hat{\Delta}_t^2|N,l,\Lambda,\Sigma\rangle = N + \frac{3}{2}\,,\\
&\langle N,l-2|\rho^2|N,l\rangle = \sqrt{(N-l+2)(N+l+2)}\,,\\
&\langle l',\Lambda'|P_2|l,\Lambda\rangle = \sqrt{\frac{2l+1}{2l'+l}}(l\Lambda 20|l'\Lambda')(l020|l'0)\,,\\
&\langle N,l|\hat{\mathbf{l}}^2\,|N,l\rangle = l(l+1)\,.
\end{aligned} \tag{6.61}$$

An example of the eigenenergies of deformed Nilsson Hamiltonian is shown in Fig. 6.5. The parity ($\pi = (-1)^l$) and the z-component of the total angular momentum ($j_z = m$) label each single-particle level . The values of the orbital ($l = 0,1,2,3\ldots$, i.e., s, p, d, ...) and the total ($j = 1/2, 3/2,\ldots$) angular momenta of the spherical nucleus are marked at $\varepsilon = 0$. The magic numbers (2, 8, 20, ...) corresponding to the closed, orbitals are also shown in the plot.

6.2.3 Higher multipolarity deformations

Deformations of higher multipolarities ε_λ are included in the oscillator part of Nilsson's Hamiltonian in a way that conserves the volume of each equipotential surface. One adds to the oscillator part a linear combination of Legendre polynomials with coefficients $2\varepsilon_\lambda$

$$\hat{H}_{\mathrm{sp}} = \frac{1}{2}\hbar\omega_0(\{\varepsilon_\lambda\})\left\{-\hat{\Delta}_t + \rho^2[1 - \frac{2}{3}\varepsilon P_2(\cos\vartheta_t) + \sum_{\lambda=1,\,\lambda\neq 2}^{\infty} 2\varepsilon_\lambda P_\lambda(\cos\vartheta_t)]\right\} + V^{\mathrm{t}}_{\mathrm{corr}}\,, \tag{6.62}$$

where $\{\varepsilon_\lambda\} \equiv \varepsilon_1, \varepsilon, \varepsilon_3, \varepsilon_4, \ldots$.
One demands that the volume $\mathscr{V}(\{\varepsilon_\lambda\})$ of the equipotential surface:

$$V_{\mathrm{osc}} = \frac{1}{2}\hbar\omega_0(\{\varepsilon_\lambda\})\rho^2\left[1 - \frac{2}{3}\varepsilon P_2(\cos\vartheta_t) + 2\sum_{\lambda=1,\lambda\neq 2}^{\infty} \varepsilon_\lambda P_\lambda(\cos\vartheta_t)\right] \tag{6.63}$$

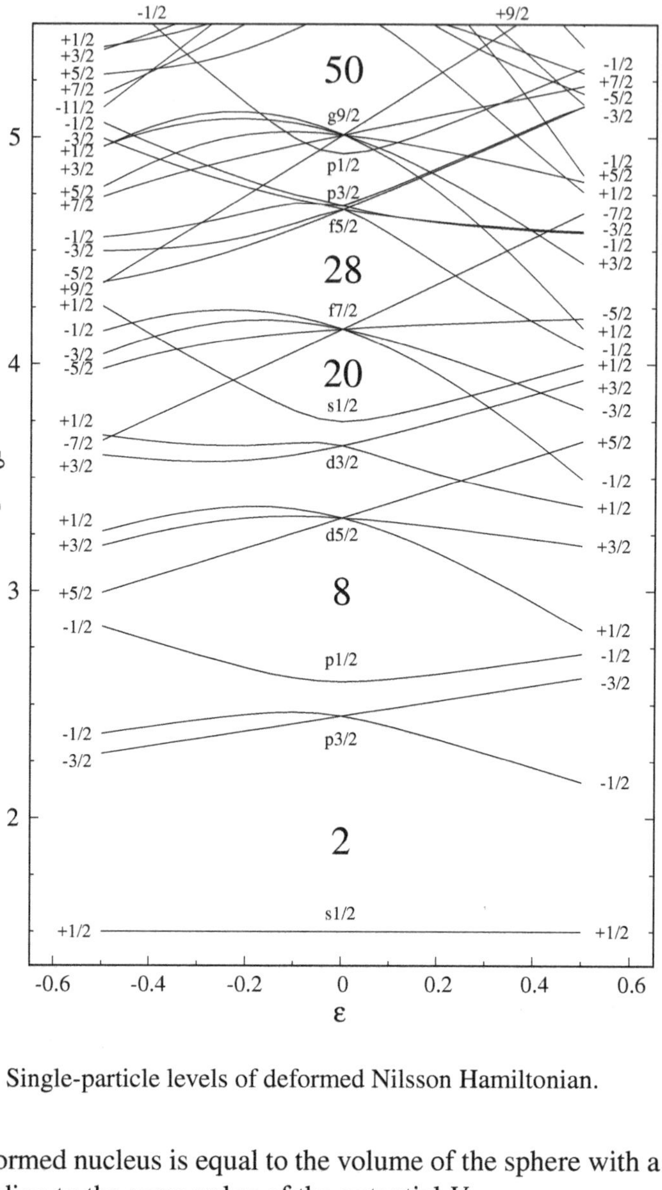

Figure 6.5 Single-particle levels of deformed Nilsson Hamiltonian.

of the deformed nucleus is equal to the volume of the sphere with a radius R_0 corresponding to the same value of the potential $V_{\rm osc}$

$$\frac{4}{3}\pi R_0^3 = \int_{\mathscr{V}(\{\varepsilon_\lambda\})} d^3r = \left[\frac{\hbar}{M\omega_0(\{\varepsilon_\lambda\})}\right]^{\frac{3}{2}} \frac{1}{(1+\frac{1}{3}\varepsilon)(1-\frac{2}{3}\varepsilon)^{\frac{1}{2}}} \int_0^{2\pi} d\varphi \int_{-1}^{1} d\cos\vartheta_t \frac{\rho^3}{3}. \tag{6.64}$$

Here, one has changed the ordinary coordinates (x, y, x) into the stretched coordinates (ξ, η, ζ) when evaluating the volume of the nucleus.

From Eq. (6.63) one gets

$$\rho^2 = \frac{2V_{\text{osc}}}{\hbar\omega_0}\left[1 - \frac{2}{3}\varepsilon P_2(\cos\vartheta_t) + 2\sum_{\lambda=1,\lambda\neq 2}^{\infty} \varepsilon_\lambda P_\lambda(\cos\vartheta_t)\right]^{-1} \tag{6.65}$$

and by inserting this result into the integral (6.64), one obtains the condition for the volume conservation in the following form:

$$\begin{aligned}\frac{4}{3}\pi R_0^3 &= \left(\frac{\hbar}{M\omega_0}\cdot\frac{2V_{\text{osc}}}{\hbar\omega_0}\right)^{3/2} \frac{2\pi}{3\left(1+\frac{1}{3}\varepsilon\right)\left(1-\frac{2}{3}\varepsilon\right)^{1/2}} \\ &\cdot \int_{-1}^{1}\left[1 - \frac{2}{3}\varepsilon P_2(\cos\vartheta_t) + 2\sum_{\substack{\lambda=1\\ \lambda\neq 2}}^{\infty} \varepsilon_\lambda P_\lambda(\cos\vartheta_t)\right]^{-3/2} d\cos\vartheta_t\,.\end{aligned} \tag{6.66}$$

For $\{\varepsilon_\lambda\} \to 0$ one has

$$\frac{4}{3}\pi R_0^3 = \left(\frac{\hbar}{M\overset{\circ}{\omega}_0}\right)^{3/2}\left(\frac{2V_{\text{osc}}}{\hbar\overset{\circ}{\omega}_0}\right)^{3/2}\cdot\frac{2\pi}{3}\cdot 2. \tag{6.67}$$

The value V_{osc} can be obtained from the last equation

$$(2V_{\text{osc}})^{3/2} = R_0^3\,\overset{\circ}{\omega}{}_0^3\,M^{3/2}\,. \tag{6.68}$$

Inserting the above relation to Eq. (6.66), one obtains the deformation dependence of the harmonic oscillator frequency needed to conserve the volume of the equipotential surfaces:

$$\frac{\omega_0(\{\varepsilon_\lambda\})}{\overset{\circ}{\omega}_0} = \left\{\int_0^1 \frac{(1+\varepsilon/3)^{-1}(1-2\varepsilon/3)^{-1/2}\,dx}{\left[1 - \frac{2}{3}\varepsilon P_2(x) + 2\sum_{\lambda=1,\lambda\neq 2}^{\infty}\varepsilon_\lambda P_\lambda(x)\right]^{3/2}}\right\}^{1/3} \tag{6.69}$$

This equation, a consequence of nuclear matter's incompressibility, forces the oscillator frequency to change when the Nilsson potential deforms.

A generalization of the Nilsson potential into the non-axial case is shown in Sec. 13.2, in which the generalized Bohr Hamiltonian is presented.

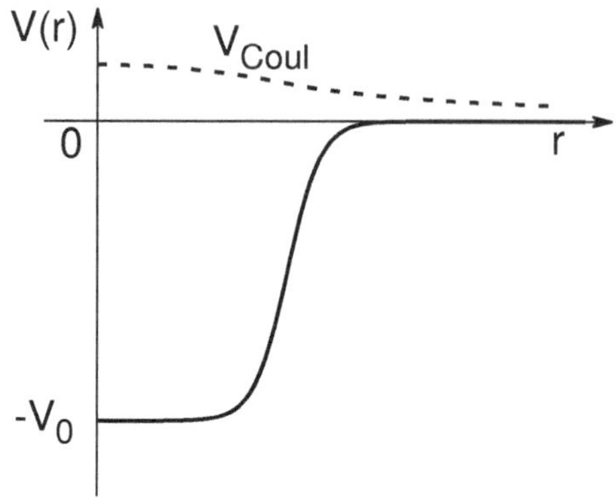

Figure 6.6 Woods-Saxon and Coulomb potentials.

6.3 WOODS-SAXON POTENTIAL

More realistic than the Nilsson potential is the mean-field potential proposed by Woods and Saxon. It is a potential well with a flat bottom and diffused walls, schematically shown in Fig. 6.6.

For the spherical nucleus, it is described by the formula:

$$V(r) = -V_{\rm WS}\left[1 + e^{\frac{r-R_0}{a}}\right]^{-1}, \tag{6.70}$$

where typically, the following parameters are assumed:

$$R_0 = 1.27A^{1/3}\ \text{fm},\ \ a = 0.67\ \text{fm}, \tag{6.71}$$

and depth

$$V_{\rm WS} = \left(51 \pm 33\frac{N-Z}{A}\right)\text{MeV}, \tag{6.72}$$

where the plus sign holds fo protons and the minus sign for neutrons.
The spin-orbit orbit potential is proportional to the derivative of the central potential (6.70)

$$\hat{V}_{\rm LS} = V_{\rm ls} R_0^2 \frac{1}{r}\frac{dV}{dr}\hat{\mathbf{l}}\cdot\hat{\mathbf{s}}, \tag{6.73}$$

where $V_{\rm ls} = -0.44$ MeV.

The Coulomb potential energy for protons is taken in as that generated by the uniform charge distribution in the nucleus:

$$V_{\text{Coul}} = \begin{cases} \dfrac{Ze^2}{r} & \text{for } r > R_0\,, \\[2ex] \dfrac{Ze^2 r^2}{R_0^3} & \text{for } r < R_0\,. \end{cases} \tag{6.74}$$

The Hamiltonian equation with Woods-Saxon potential one solves numerically. Deformation is introduced into the Woods-Saxon potential, making the potential radius R_0 dependent on the angle (5.44), similarly as it is done in the Nilsson's potential, and keeping the constant surface diffusion width.

6.4 YUKAWA FOLDED MEAN-FIELD POTENTIAL

Effective forces may be used to evaluate an average mean-field potential in atomic nuclei. The so-called Yukawa-folded potential is obtained by integrating a product of the two-body Yukawa forces (2.4) and the uniform (in better approximations diffused) density distribution in the deformed nucleus

$$V(\mathbf{r}) = -\frac{V_0}{4\pi a^3}\int_{\mathcal{V}} \frac{e^{-|\mathbf{r}-\mathbf{r}'|/a}}{|\mathbf{r}-\mathbf{r}'|/a}\, d^3r'\,, \tag{6.75}$$

were adjusted to the experimental data, $V_0 = 52.5$ MeV is the depth of the potential well, and the parameter $a = 0.8$ fm is the range of the Yukawa function.

The Yukawa folded potential describes spherical and deformed nuclei well, even at very elongated shapes that appear close to the scission configuration. Contrary to the Woods-Saxon potential, its surface thickness is not constant.

The spin-orbit potential is assumed to be proportional to the gradient of the central potential (6.75)

$$\hat{V}_{\text{LS}}(\mathbf{r}) = -\lambda\left(\frac{\hbar}{2Mc}\right)^2 \hat{\sigma}\cdot(\nabla V\cdot\mathbf{p}/\hbar)\,, \tag{6.76}$$

where λ is the strength of the spin-orbit interaction and is given by the formulae

$$\lambda_{\text{p}} = 28.0 + 6.0\frac{A}{240} \quad \text{and} \quad \lambda_{\text{n}} = 31.5 + 4.5\frac{A}{240}$$

for protons and neutrons, respectively. The Coulomb potential, which has to be added for protons, is given by the integral

$$\hat{V}_{\rm Coul}(\mathbf{r}) = \frac{3Ze^2}{4\pi R_0^3} \int_{\mathscr{V}} \frac{d^3r'}{|\mathbf{r}-\mathbf{r}'|} . \tag{6.77}$$

The radial dependence of the Yukawa-folded potential is similar to that of Woods-Saxon's. Its main advantage is the better consistency of the deformed mean-field potential and the density of the nucleus than in the case of the Nilsson or the Woods-Saxon potentials. At best, such consistency of the mean-field potential and density is achieved in the Hartree-Fock method, also called the self-consistent field method described in Chapter 7.

When applying the micro-macro model to practical applications, the problem of finding an effective and flexible shape parametrization of the deformed single-particle potential well and the corresponding macroscopic density distribution appears. This problem is especially difficult when one would like to have the same shape parametrization both in the ground state and around the scission configuration when nucleus fission occurs. The following section will present a model that fulfills both requirements, is rapidly convergent, and is easy to implement.

6.5 FOURIER SHAPE PARAMETRIZATION

The enormous variety of shapes encountered all across the nuclear chart, between the oblate side found in the transition region, corresponding to the progressive filling of the pf shell, and the prolate deformations realized in the rare-earth and actinide regions, or even more so for the very elongated and necked-in shapes met in the course of the fission process, has practically been an everlasting subject of intensive research and modeling since the early days of nuclear physics. Good shape parametrizations should be flexible and close to the physical reality, but they must also involve as few deformation parameters as possible. As an example of such a parametrization, let us mention the expansion of the nuclear radius in spherical harmonics proposed by Lord Rayleigh in the 19th century (confer Eq. 5.44). It is the most widely used shape parametrization in nuclear structure calculations. It has, however, been shown that a vast number of deformation parameters $a_{\lambda\mu}$ (5.44) would be needed (up to 8th order) to reproduce the fission barrier heights correctly in the actinide region, already when restricting oneself to axial and left-right symmetry. It would imply that one must handle functions of several dimensions when searching for stationary points in a multi-dimensional deformation space to locate ground state, isomeric states, saddle points, valleys, and ridges between them.

Many other shape parametrizations have been proposed since the days of Lord Rayleigh, among which one needs to mention the so-called *quadratic surfaces of revolution*, the Cassini ovals, the so-called *Funny-Hills* shapes, as well as an expansion of the nuclear surface in a series of Legendre polynomials in the cylindrical coordinates. Some of these shape parametrizations can describe nuclear potential-energy surfaces rather well with a limited number of collective variables but have the fundamental inconvenience that there is no way of controlling their convergence. This disadvantage is easy to avoid in a very efficient parametrization obtained by the Fourier expansion of the nuclear surface. This expansion rapidly converges and accurately describes shapes around the ground state of nuclei and very elongated shapes that appear during the fission process up to the scission configuration. It is very convenient to use the cylindrical set of coordinates (ρ, φ, z) to describe axially symmetric shapes (a generalization to triaxial shapes will be given below). Let us introduce the distance ρ_s of a surface point at coordinate z from the symmetry axis, which coincides with the z-axis. The square of this distance can be expanded in the Fourier series

$$\frac{\rho_s^2(z)}{R_0^2} = \sum_{n=1}^{\infty} \left[a_{2n} \cos\left(\frac{(2n-1)\pi}{2} \frac{z - z_{sh}}{z_0} \right) + a_{2n+1} \sin\left(\frac{2n\pi}{2} \frac{z - z_{sh}}{z_0} \right) \right] , \tag{6.78}$$

where R_0 is the radius of the corresponding spherical shape having the same volume. The extension of the shape along the symmetry axis is $2z_0 \equiv 2cR_0$ with left and right ends located at $z_{\rm min} = z_{\rm sh} - z_0$ and $z_{\rm max} = z_{\rm sh} + z_0$, where $\rho_s^2(z)$ vanishes, a condition which is automatically satisfied by Eq. (6.78). Here, $c = z_0/R_0$ is the Funny-Hills elongation parameter ($c < 1$ for oblate, $c > 1$ for prolate shapes), which is related to the even Fourier deformation parameters by the volume-conservation condition:

$$\frac{\pi}{3c} = \sum_{n=1}^{\infty} (-1)^{n-1} \frac{a_{2n}}{2n-1} . \tag{6.79}$$

The Fourier expansion coefficients a_i are treated in the following as the deformation parameters. An example of a nuclear shape generated that might be realized is displayed in Fig. 6.7, where all variables that appear in Eq. (6.78) are explained.

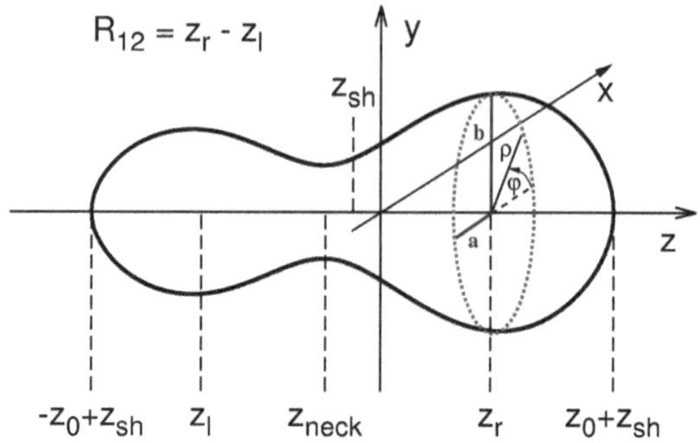

Figure 6.7 Schematic visualization in cylindrical coordinates of the parameters entering the definition of the profile function defined with Eq. (6.78). The coordinates z_l and z_r correspond to the mass centers of the left and right nascent fragment, respectively.

The shift $z_{\rm sh}$ in (6.78) is chosen to keep the center of mass of the nuclear in the origin of the coordinate system. The distance between the mass centers of both fragments is $R_{12} = z_{\rm r} - z_{\rm l}$. The parameters a_2, a_3, a_4 describe quadrupole, octupole, and hexadecapole type deformations, which are also related to elongation, left-right asymmetry, and neck thickness, respectively.

To describe non-axial shapes, one assumes that the cross-section of the nucleus perpendicular to the symmetry axis z is an ellipsoid, which half-axes a and b define the following non-axial deformation parameter

$$\eta = \frac{b-a}{b+a} \,. \tag{6.80}$$

Assuming that this parameter is the same at all z-constant crosses of nuclear surface, the profile function can be written in cylindrical coordinates in the form

$$\varrho_s^2(z,\varphi) = \rho_s^2(z)\,\frac{1-\eta^2}{1+\eta^2+2\eta\cos(2\varphi)} \,, \tag{6.81}$$

with $\rho_s^2(z)$ given by Eq. (6.78).

The above-defined shape parametrization is rapidly converging, even for strongly deformed shapes. One inconvenience of the definition of the Fourier coefficients a_n is, however, that, contrary to the Funny-Hills elongation parameter c, the Fourier coefficient a_2 decreases with increasing elongation of the nucleus. One introduces the following new collective coordinates to cure this inconvenience:

$$
\begin{aligned}
q_2 &= \frac{a_2^{(0)}}{a_2} - \frac{a_2}{a_2^{(0)}} \,, \\
q_3 &= a_3 \,, \\
q_4 &= a_4 + \sqrt{\left(\tfrac{q_2}{9}\right)^2 + \left(a_4^{(0)}\right)^2} \,, \\
q_5 &= a_5 - (q_2 - 2)\tfrac{a_3}{10} \,, \\
q_6 &= a_6 - \sqrt{\left(\tfrac{q_2}{100}\right)^2 + \left(a_6^{(0)}\right)^2} \,,
\end{aligned}
\tag{6.82}
$$

where the $a_n^{(0)}$ are the values of the Fourier coefficients for the sphere. In addition, these new collective q_n coordinates are equal to zero for spherical shapes.

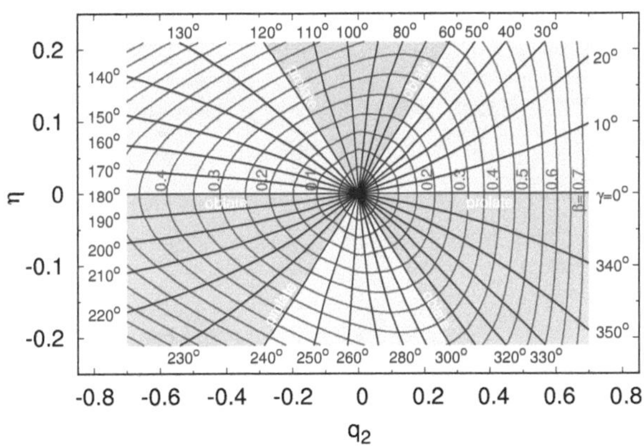

Figure 6.8 Visualization of the relation between (β, γ) and (q_2, η) collective coordinates.

It would be good to establish the connection between the above collective coordinates (q_2, η) and the frequently used (β, γ) deformation parameters (confer also Eq. 5.44) defined by Bohr as

$$
\beta = \frac{1}{X}\sqrt{Q_{20}^2 + Q_{22}^2} \qquad \text{and} \qquad \gamma = \arctan\left(\frac{Q_{22}}{Q_{20}}\right)
\tag{6.83}
$$

where the constant $X = 3\,r_0^2 A^{5/3} / \sqrt{5\pi}$ and Q_{20}^2 and Q_{22}^2 are the components of the mass quadrupole tensor

$$
Q_{20} = \langle 2z^2 - x^2 - y^2 \rangle, \quad Q_{22} = \langle y^2 - x^2 \rangle \,.
\tag{6.84}
$$

This relation between (q_2, η) and (β, γ) is shown in Fig. 6.8, where, of course, the well known γ-deformation 60° symmetry is visible.

Due to this symmetry, one needs, however, to be very cautious when investigating deformation-energy landscapes in the (η, q_2, q_3, q_4) deformation space to avoid any ‘double-counting’ of similar but differently oriented shapes. From Fig. 6.8 one immediately concludes that the shape realized by e.g. ($q_2 \approx +0.17, \eta \approx 0.13$) is equivalent to the one at ($\beta = 0.25, \gamma = 60°$). Such a comparison can, of course, only be carried out when restricted to deformed spheroidal shapes only. When higher-order multipolarities come into play, such simple correspondence is partly distorted. Please also notice that a constant value of β does not correspond to a constant elongation of the shape, as is visible in Figure 6.8.

CHAPTER 7

Self-consistent methods and density functionals

As mentioned in the introduction, a quantum mechanical problem for many bodies is not exactly solvable for the number of particles greater than two. Therefore, approximate methods must be used. The method proposed by Hartree and Fock considers the influence of remaining particles on a selected single particle with average potential. The average potential field generated by surrounding particles is determined by an iterative method to reconcile them. The total energy of the nucleus may be written as an integral over space of some energy density. In particular cases, like some versions of the Skyrme forces, the energy density derived from a zero-range pseudo-potential depends on local densities and their derivatives. In other cases, e.g., Gogny forces or the relativistic mean-field theory (RMFT), the energy density functional (EDF) has a more complex structure.

Below, the process of obtaining such an approximate solution to the many-body problem will be discussed in detail. First, let us recall the second quantization formalism for fermions, in which Hartree-Fock equations are written especially simply.

7.1 SECOND QUANTIZATION FORMALISM

We have already met the formalism of second quantization for bosons by solving the eigenproblem of the Hamiltonian with harmonic oscillator potential. It is based on the creation and annihilation operators of oscillation quanta or particles with total spin equal to zero. Usually, such creation and annihilation operators are defined as appropriate combinations of position and momentum coordinates to fulfill the commutation conditions for boson operators (6.3). In the representation of the number of occupations in the vacuum of bosons, one can create any number of particles and the function of their system is symmetrical due to the exchange of the pair of particles.

Fermions, particles with a half-spin, must obey the Pauli principle: only one particle can occupy every quantum state. This principle ensures the antisymmetry of fermions' system function due to particle transposition and corresponding anti-commutation conditions for fermions' creation and annihilation operators.

Consider the system of single-fermion states $\{\nu\}$:

$$\{\nu\} = \varphi_{\nu 1}, \varphi_{\nu 2}, \varphi_{\nu 3}, \ldots, \varphi_{\nu N} \,. \tag{7.1}$$

These functions are denoted by a set of quantum numbers ν_i defining clearly the φ_{ν_i} status. The antisymmetric function of N fermions can be written as the Slater determinant (3.3)

$$\psi(\mathbf{x}_1, \mathbf{x}_2, \ldots, \mathbf{x}_N) = \frac{1}{\sqrt{N!}} \begin{vmatrix} \varphi_{\nu_1}(\mathbf{x}_1) & \varphi_{\nu_1}(\mathbf{x}_2)\ldots & \varphi_{\nu_1}(\mathbf{x}_N) \\ \varphi_{\nu_2}(\mathbf{x}_1) & \varphi_{\nu_2}(\mathbf{x}_2)\ldots & \varphi_{\nu_2}(\mathbf{x}_N) \\ \cdot & \cdot & \cdot \\ \cdot & \cdot & \cdot \\ \cdot & \cdot & \cdot \\ \varphi_{\nu_N}(\mathbf{x}_1) & \varphi_{\nu_N}(\mathbf{x}_2)\ldots & \varphi_{\nu_N}(\mathbf{x}_N) \end{vmatrix} , \tag{7.2}$$

where $\mathbf{x}_i$ stays for the space, spin, and isospin coordinates of the i-th nucleon.

In the image of the second quantization, this function is presented as a result of operators a_ν^+ creating fermions in a vacuum state of particles $|0\rangle$

$$\psi = a_{\nu 1}^+ a_{\nu 2}^+ \cdots a_{\nu N}^+ |0\rangle \equiv |\{\nu\}\rangle \,. \tag{7.3}$$

The vacuum state must be normalized

$$\langle 0|0\rangle = 1 \tag{7.4}$$

and one cannot annihilate a fermion here

$$a_\nu |0\rangle = 0 \,. \tag{7.5}$$

If one rearranges a pair of particles in Eq. (7.2), the determinant will change the sign

$$\psi(\mathbf{x}_1, \mathbf{x}_2 \cdots \mathbf{x}_N) = -\psi(\mathbf{x}_2, \mathbf{x}_1, \cdots, \mathbf{x}_N) \,,$$

and thus changing the order of the operators a_ν^+ in (7.3) we must change the sign of the function

$$a_{\nu 1}^+ a_{\nu 2}^+ \cdots a_{\nu N}^+ |0\rangle = -a_{\nu 2}^+ a_{\nu 1}^+ \cdots a_{\nu N}^+ |0\rangle \,. \tag{7.6}$$

The resulting anti-commutation conditions for creation operators of fermions are:

$$\{a^+_{\nu_1}, a^+_{\nu_2}\} = a^+_{\nu_1}a^+_{\nu_2} + a^+_{\nu_2}a^+_{\nu_1} = 0\,, \tag{7.7}$$

which ensures that Pauli's rule is fulfilled because

$$2a^+_\nu a^+_\nu = 0\,, \tag{7.8}$$

i.e., one cannot create two fermions in one state ν.

One obtains an annihilation operator by taking the Hermitian conjugation of the creation operator

$$a_\nu = (a^+_\nu)^+\,, \tag{7.9}$$

$$\{a_{\nu_1}, a_{\nu_2}\} = 0\,. \tag{7.10}$$

The orthogonality condition of the many-body wave functions:

$$\int \psi^\star(\mathbf{x}_{\nu_1}, \mathbf{x}_{\nu_2}, \ldots, \mathbf{x}_{\nu_N})\,\psi(\mathbf{x}_{\mu_1}, \mathbf{x}_{\mu_2}, \ldots \mathbf{x}_{\mu_N})d\tau_1 d\tau_2 \cdots d\tau_N$$

$$= \langle 0|a_{\nu_N}a_{\nu_{N-1}} \cdots a_{\nu_1}a^+_{\mu_1}a^+_{\mu_2} \cdots a^+_{\mu_N}|0\rangle \tag{7.11}$$

$$= \begin{cases} -1 & \text{for } \nu \text{ and } \mu \text{ differing by odd permutation number,} \\ 0 & \text{if } \nu \text{ and } \mu \text{ differ by at least one quantum number,} \\ 1 & \text{for } \nu \text{ and } \mu \text{ differing by even permutation number,} \end{cases}$$

implies the following anti-commutation condition of fermion creation and annihilation operators:

$$\{a_\nu, a^+_\mu\} = \delta_{\mu\nu}\,. \tag{7.12}$$

If one wants to use many-body states in the representation of the number of occupations, one must also use the second quantization form $\hat{O}^{\mathrm{II}}$ of the fermion operators $\hat{O}$, i.e., one has to express them with fermions creation and annihilation operators in such a way that their mean values in the state (7.3) and (7.2) are the same

$$\int \psi^\star \hat{O}\psi d\tau_1 \cdots d\tau_N = \langle\{\nu\}|\hat{O}^{\mathrm{II}}|\{\nu\}\rangle\,. \tag{7.13}$$

The particle number operator will count the states occupied by fermions

$$\hat{N} = \sum_\nu a^+_\nu a_\nu \tag{7.14}$$

and it is easy to show for the function ψ (7.3) that it has eigenvalue equal to N:

$$\hat{N}|\psi\rangle = \sum_{\nu} a_{\nu}^{+} a_{\nu} \{a_{\nu_1}^{+} \cdots a_{\nu_N}^{+} |0\rangle\} = N|\psi\rangle . \tag{7.15}$$

One-body operator, e.g., kinetic energy $\hat{T}$, is the sum of single-particle operators $\hat{t}(i)$ with matrix elements $\langle\mu|\hat{t}|\nu\rangle$

$$\hat{T} = \sum_{i=1}^{N} \hat{t}(i) . \tag{7.16}$$

The average value of the $\hat{T}$ operator in the ψ state described by the formula (7.2) is

$$\begin{aligned} \langle\psi|\hat{T}|\psi\rangle &= \int \psi(\mathbf{x}_1, \mathbf{x}_2, \ldots, \mathbf{x}_N)^{\star} \left(\sum_{i} \hat{t}(i)\right) \psi(\mathbf{x}_1, \mathbf{x}_2, \ldots, \mathbf{x}_N) d\tau_1 \cdots d\tau_N \\ &= \sum_{\nu} \langle\nu|\hat{t}|\nu\rangle , \end{aligned} \tag{7.17}$$

and is the same as in the occupation number representation (7.3), if one will present the $\hat{T}$ operator in the form

$$\hat{T}^{\mathrm{II}} = \sum_{\mu\nu} \langle\mu|\hat{t}|\nu\rangle a_{\mu}^{+} a_{\nu} . \tag{7.18}$$

One can check this by a direct calculation

$$\begin{aligned} \langle\psi|\hat{T}^{\mathrm{II}}|\psi\rangle &= \langle 0|a_{\nu_N} \cdots a_{\nu_1}| \sum_{\mu\nu} \langle\mu|\hat{t}|\nu\rangle a_{\mu}^{+} a_{\nu} |a_{\nu_1}^{+} \cdots a_{\nu_N}^{+}|0\rangle \\ &= \sum_{\mu\nu} \langle\mu|\hat{t}|\nu\rangle \langle 0| \prod_{\mu\nu} (\delta_{\mu\nu} - a_{\mu}^{+} a_{\nu})|0\rangle \\ &= \sum_{\nu} \langle\nu|\hat{t}|\nu\rangle . \end{aligned} \tag{7.19}$$

Two-body operator

$$\hat{V} = \sum_{i<j} \hat{v}(i,j) \tag{7.20}$$

will be presented in the second quantization in the form

$$\hat{V}^{\mathrm{II}} = \frac{1}{2} \sum_{\mu\nu\gamma\delta} \langle\mu\nu|\hat{v}|\gamma\delta\rangle a_{\mu}^{+} a_{\nu}^{+} a_{\delta} a_{\gamma} , \tag{7.21}$$

where $\mu \neq \nu$ and $\gamma \neq \delta$.
The average value of $\hat{V}^{\mathrm{II}}$ in the state ψ (7.3) is

$$\begin{aligned}\langle\psi|\hat{V}^{\mathrm{II}}|\psi\rangle &= \langle 0|a_{\beta_N}\cdots a_{\beta_1}\frac{1}{2}\sum_{\mu\nu\gamma\delta}\langle\mu\nu|\hat{v}|\gamma\delta\rangle a^+_\mu a^+_\nu a_\delta a_\gamma|a^+_{\beta_1}\cdots a^+_{\beta_N}|0\rangle \\ &= \frac{1}{2}\sum_{\mu\nu}\langle\mu\nu|\hat{v}|\gamma\delta\rangle(\delta_{\gamma\mu}\delta_{\delta\nu}-\delta_{\gamma\nu}\delta_{\delta\mu}) = \\ &= \frac{1}{2}\sum_{\mu\nu}(\langle\mu\nu|\hat{v}|\mu\nu\rangle - \langle\mu\nu|\hat{v}|\nu\mu\rangle)\,.\end{aligned} \tag{7.22}$$

We used here the conditions of anti-commutation of fermion operators. We will get the same average value by integrating the operator $\hat{V}$ (7.21) between functions (7.2)

$$\begin{aligned}&\int\psi^\star\hat{V}\psi d\tau_1\cdots d\tau_N = \\ &= \int\psi^\star(\mathbf{x}_1,\mathbf{x}_2,\ldots,\mathbf{x}_N)\sum_{i<j}\hat{v}(i,j)\psi(\mathbf{x}_1,\mathbf{x}_2,\ldots,\mathbf{x}_N)d\tau_1\cdots d\tau_N = \\ &= \frac{1}{2}\sum_{\mu\nu}\left[\int\varphi^\star_\mu(\mathbf{x}_i)\varphi^\star_\nu(\mathbf{x}_j)\,\hat{v}(i,j)\left(\varphi_\mu(\mathbf{x}_i)\varphi_\nu(\mathbf{x}_j)-\varphi_\nu(\mathbf{x}_i)\varphi_\mu(\mathbf{x}_j)\right)d\tau_i d\tau_j\right].\end{aligned} \tag{7.23}$$

The last term in this equation describes the exchange effects that appear due to the antisymmetry of the wave function. This element is omitted in the Hartree method . In particular, the Hamiltonian of N particles often consists of the sum of single-body Hamiltonians $\hat{h}$ and double-body interactions $\hat{v}$

$$\hat{H} = \sum_{i=1}^{N}\hat{h}(i) + \sum_{i\neq j}\hat{v}(i,j)\,. \tag{7.24}$$

In the second quantization language, it will take the form

$$\hat{H}^{\mathrm{II}} = \sum_{\mu\nu}\langle\mu|\hat{h}|\nu\rangle a^+_\mu a_\nu + \frac{1}{2}\sum_{\mu\nu\gamma\delta}\langle\mu\nu|\hat{v}|\gamma\delta\rangle a^+_\mu a^+_\nu a_\delta a_\gamma\,. \tag{7.25}$$

If the $|\nu\rangle$ states are the eigenfunctions of a single-particle Hamiltonian:

$$\hat{h}|\nu\rangle = e_\nu|\nu\rangle\,, \tag{7.26}$$

then Hamiltonian $\hat{H}^{\mathrm{II}}$ will be written particularly simply

$$\hat{H}^{\mathrm{II}} = \sum_{\nu} e_\nu a^+_\nu a_\nu + \frac{1}{2}\sum_{\mu\nu\gamma\delta}\langle\mu\nu|\hat{v}|\gamma\delta\rangle a^+_\mu a^+_\nu a_\delta a_\gamma\,. \tag{7.27}$$

7.2 HARTREE-FOCK METHOD

In the shell model, one assumes independent nucleon motion in a mean-field described by the phenomenological Nilsson or Woods-Saxon potentials. This single-particle field should be consistent with the effective one built from nucleon-nucleon interaction. Hartree-Fock theory allows to generate the average potential field from two-body forces and thus obtain a single-particle Hamiltonian containing information about two-particle nuclear interactions. Since one does not know precisely the form of effective two-body interactions, one can only verify them by comparing measurable properties of the nucleus to average values of relevant operators calculated between Hartree-Fock functions.

Consider the Hamiltonian (7.25) with effective two-body interaction $\hat{v}(i,j)$. Our goal will be to find such an operator of single-particle mean-field potential U that will satisfy the condition

$$\langle r|U|s\rangle = \sum_{m=1}^{N} (\langle rm|\hat{v}|sm\rangle - \langle rm|\hat{v}|ms\rangle) . \tag{7.28}$$

Let $|l\rangle$, $|m\rangle$, $|n\rangle, \ldots, |r\rangle$, $|s\rangle$ denote states occupied by particles, and $|a\rangle, |b\rangle, |c\rangle, \ldots$ empty states, and they will be eigenfunctions of a single-particle Hamiltonian

$$\hat{h}_{\mathrm{sp}} = \hat{t} + U , \tag{7.29}$$

which, in the image of the second quantization, has the form

$$\hat{H}^{\mathrm{II}}_{\mathrm{sp}} = \sum_{ki} \left\{ \langle k|\hat{t}|i\rangle + \sum_{m=1}^{N} (\langle km|\hat{v}|im\rangle - \langle km|\hat{v}|mi\rangle) \right\} a^{+}_{k} a_{i} . \tag{7.30}$$

The U field can be found by following iterations:

1. One assumes the sample of trial functions $|k\rangle, \ldots$ occupied by particles,
2. From Eq. (7.28), one can find the potential U,
3. Then one solves the eigenproblem of $\hat{h}_{\mathrm{sp}}$

   $$(\hat{t} + U)|l\rangle = e_l |l\rangle , \tag{7.31}$$

   to get the new $|l\rangle$ function set,
4. One compares them with previous functions $|k\rangle$ and if one has not obtained self-consistence, one repeats the procedure from the first point, assuming the new states $|l\rangle$ for $|k\rangle$.

In this way, one is passing from two-body (7.25) Hamiltonian to the new Hamiltonian $\hat{H}$ with the dominant single-body part:

$$\hat{H} = \sum_i (\hat{t}(i) + U(i)) + \left(\sum_{i<j} \hat{v}(i,j) - \sum_i U(i) \right). \tag{7.32}$$

The approximation has been made that the contribution to the ground state energy, derived from the so-called residual interactions: $V_{\text{res}} = V - U$, is small. In Chapters 8 and 9, additional methods will be discussed that take into account residual interactions such as long- and short-range forces.

The Hartree-Fock self-consistent solution reached through subsequent iterations is equivalent to the variational principle

$$\delta \frac{\langle \psi_0 | \hat{H} | \psi_0 \rangle}{\langle \psi_0 | \psi_0 \rangle} = 0, \tag{7.33}$$

where ψ_0 is the antisymmetric product (7.2) of the single particle functions

$$|\psi_0\rangle = \prod_{m=1}^{N} a_m^+ |0\rangle \tag{7.34}$$

and variation means taking into account particle-hole excitations

$$\delta |\psi_0\rangle \sim a_b^+ a_m |\psi_0\rangle. \tag{7.35}$$

Proper selection of the N involved states ensures an equation equivalent to the condition (7.28)

$$\langle m | \hat{t} | b \rangle + \sum_{n=1}^{N} (\langle mn | \hat{v} | bn \rangle - \langle nm | \hat{v} | bn \rangle) = 0. \tag{7.36}$$

Using this equation and the definition of single-particle Hamiltonian $\hat{h}_{\text{sp}}$ (7.28)-(7.29) one gets

$$\langle m | \hat{h}_{\text{sp}} | b \rangle = 0. \tag{7.37}$$

It follows that matrix elements between the occupied $|m\rangle$ and empty $|b\rangle$ states disappear, while the Hamiltonian single-particle equation is met (7.31)

$$\langle l | \hat{h}_{\text{sp}} | m \rangle = e_l \delta_{lm}. \tag{7.38}$$

The Hartree-Fock method is widely used to evaluate approximate wave functions and energies of nuclei using the effective forces of, e.g., Gogny, Skyrme, or RMFT types.

7.3 GOGNY INTERACTION

The Gogny density-dependent effective nucleon-nucleon force is of the following form:

$$
\begin{aligned}
V_{12} &= \sum_{i=1}^{2} (W_i + B_i\hat{P}_\sigma - H_i\hat{P}_\tau - M_i\hat{P}_\sigma\hat{P}_\tau)\, e^{-\frac{(\vec{r}_1-\vec{r}_2)^2}{\mu_i^2}} \\
&+ i\, W_{\mathrm{LS}}\, (\overleftarrow{\nabla_1 - \nabla_2}) \times \delta(\vec{r}_1 - \vec{r}_2)(\overrightarrow{\nabla_1 - \nabla_2}) \cdot (\vec{\sigma}_1 + \vec{\sigma}_2) \qquad (7.39) \\
&+ t_0\, (1 + x_0\hat{P}_\sigma)\, \delta(\vec{r}_1 - \vec{r}_2) \left[\rho(\frac{\vec{r}_1 - \vec{r}_2}{2})\right]^\gamma + V_{\mathrm{Coul}}\,.
\end{aligned}
$$

The first three terms represent a central finite range interaction, a zero-range spin-orbit term and a zero-range density-dependent force, respectively, and the fourth one is the Coulomb interaction between protons. The central interaction is made of two distinct Gauss functions with ranges μ_1 and μ_2, whose values are given below. $\hat{P}\sigma$ and $\hat{P}\tau$ denote the spin and isospin exchange operators, respectively, and ρ represents the total density.

Performing HF calculation, one takes some effective interactions between the two nucleons placed in $\vec{r}_1$ and $\vec{r}_2$ positions surrounded by the nuclear matter, which demands keeping the Pauli principle. The starting point to find this interaction is the pure nucleon-nucleon force in a vacuum, described in Sec. 2. The nucleons in the neighborhood of other ones do not feel these pure forces, so the phenomenological effective interaction must be used, which demands applying the many-body method, like HF. The parameters of effective forces are obtained by fitting their results to the features of magic nuclei and nuclear matter.

The Gogny force consists of two central terms, depending only on the two nucleons' distance: one corresponds to the repulsive, short-range Gaussian, the second one to the attracting interaction shown in Fig. 1.1. Then come the non-central spin-orbit, density-dependent, and Coulomb terms.

The effective Gogny forces fulfill the same symmetry rules as the pure nucleon-nucleon interaction, described in Sec. 1.4, i.e., they have to be invariant with respect to:

1. Space translations,
2. Lorentz transformation,
3. Space reflections,
4. Time reversal,
5. Spin and isospin exchange,

The above symmetries imply that the central part of the Gogny potential can be constructed only from the scalar quantities: $\hat{\mathbf{r}}^2$, $\hat{\mathbf{p}}^2$, $\hat{\mathbf{s}}_1 \cdot \hat{\mathbf{s}}_2$, $\hat{\mathbf{t}}_1 \cdot \hat{\mathbf{t}}_2$, where $\hat{\mathbf{r}}$ denotes the relative distance operator, $\hat{\mathbf{p}}$ relative momentum, $\hat{\mathbf{s}}$ and $\hat{\mathbf{t}}$ are the spin and isospin operators of individual nucleons. Without momentum dependence, the central part of the potential has the following general form:

$$\hat{V}_{cent} = U(r) + U_s(r)(\hat{\mathbf{s}}_1 \cdot \hat{\mathbf{s}}_2) + U_t(r)(\hat{\mathbf{t}}_1 \cdot \hat{\mathbf{t}}_2) + U_{st}(r)(\hat{\mathbf{s}}_1 \cdot \hat{\mathbf{s}}_2)(\hat{\mathbf{t}}_1 \cdot \hat{\mathbf{t}}_2) . \quad (7.40)$$

These expressions are connected with the coordinates exchange operators of two particles, described in Subs. 2.1.6.
They are:

a) Wigner operator $\hat{P}_W$, which does not change wave function,
b) Majorana operator $\hat{P}_M$, exchanging space,
c) Bartlet operator $\hat{P}_B$, exchanging spins,
d) Heisenberg operator $\hat{P}_H$, exchanging spin and space coordinates:

$$\begin{aligned} \hat{P}_W \psi(\mathbf{r}_1, s_1; \mathbf{r}_2, s_2) &= \psi(\mathbf{r}_1, s_1; \mathbf{r}_2, s_2) , \\ \hat{P}_M \psi(\mathbf{r}_1, s_1; \mathbf{r}_2, s_2) &= \psi(\mathbf{r}_2, s_1; \mathbf{r}_1, s_2) , \\ \hat{P}_B \psi(\mathbf{r}_1, s_1; \mathbf{r}_2, s_2) &= \psi(\mathbf{r}_1, s_2; \mathbf{r}_2, s_1) , \\ \hat{P}_H \psi(\mathbf{r}_1, s_1; \mathbf{r}_2, s_2) &= \psi(\mathbf{r}_2, s_2; \mathbf{r}_1, s_1) . \end{aligned} \quad (7.41)$$

Double acting of the above operators returns wave functions to themselves

$$\hat{P}_W^2 = \hat{P}_M^2 = \hat{P}_B^2 = \hat{P}_H^2 = 1 . \quad (7.42)$$

The Bartlet operator $\hat{P}_B$ acts only on the spin part of wave function

$$\hat{P}_B \zeta(s_1, s_2) = \zeta(s_2, s_1) = p_B \zeta(s_1, s_2) . \quad (7.43)$$

Table 7.1 shows the spin wave functions of two nucleon systems in the representation of coupled moments. They are eigenstates of the square of the total spin operator:

$$\hat{S}^2 = (\hat{\mathbf{s}}_1 + \hat{\mathbf{s}}_2)^2 \quad (7.44)$$

and its third component:

$$\hat{S}_z = \hat{s}_{1z} + \hat{s}_{2z} . \quad (7.45)$$

The states $\zeta(+1/2)$ and $\zeta(-1/2)$ are spin wave functions of nucleons. The first three states in the table are functions of the triplet state and do not change

Table 7.1 Spin wave functions of the two nucleons in the coupled moment representation and the corresponding quantum numbers: total spin S and its projection S_z on z-axis.

Eigenfunction	S	S_z
$\zeta_1(-1/2)\zeta_2(-1/2)$	1	−1
$\zeta_1(+1/2)\zeta_2(+1/2)$	1	+1
$\frac{1}{\sqrt{2}}[\zeta_1(+1/2)\zeta_2(-1/2)+\zeta_1(-1/2)\zeta_2(+1/2)]$	1	0
$\frac{1}{\sqrt{2}}[\zeta_1(+1/2)\zeta_2(-1/2)-\zeta_1(-1/2)\zeta_2(+1/2)]$	0	0

during nucleon exchange; they are symmetric. The last - singlet state is antisymmetric. Acting with the Bartlet operator on these functions, one gets eigenvalues:

$$p_B = (-1)^{S+1} \,. \tag{7.46}$$

The operator with the same eigenvalues is:

$$\frac{1}{2}(1+4\hat{\mathbf{s}}_1\cdot\hat{\mathbf{s}}_2) = \frac{1}{2} + \hat{\mathbf{S}}\cdot\hat{\mathbf{S}} - \hat{\mathbf{s}}_1\cdot\hat{\mathbf{s}}_1 - \hat{\mathbf{s}}_2\cdot\hat{\mathbf{s}}_2 \,. \tag{7.47}$$

So, one can write:

$$\hat{P}_B = \frac{1}{2}(1+4\hat{\mathbf{s}}_1\cdot\hat{\mathbf{s}}_2) = \frac{1}{2}(1+\hat{\sigma}_1\cdot\hat{\sigma}_2) = \hat{P}_\sigma \,, \tag{7.48}$$

where the connection between spin operators and Pauli matrices $\hat{\sigma}$ is used.

$$\hat{\mathbf{s}} = \frac{1}{2}\hat{\sigma} \,. \tag{7.49}$$

Analogically, one can introduce the spin exchange operator :

$$\hat{P}_\tau = \frac{1}{2}(1+4\hat{\mathbf{t}}_1\cdot\hat{\mathbf{t}}_2) = \frac{1}{2}(1+\hat{\tau}_1\cdot\hat{\tau}_2) \,, \tag{7.50}$$

where $\hat{\tau}$ denote isospin Pauli matrices. The full wave function of the two nucleons, with space, spin, and isospin parts, has to be asymmetric to fulfill the Pauli principle. It means:

$$\hat{P}_M\hat{P}_\sigma\hat{P}_\tau = -1 \,, \tag{7.51}$$

and

$$\hat{P}_M = -\hat{P}_\sigma \hat{P}_\tau \, . \tag{7.52}$$

From the definition of Majorana and Heisenberg operators, one has:

$$\hat{P}_H = \hat{P}_M \hat{P}_B \, . \tag{7.53}$$

From Majorana and Bartlet operators

$$\hat{P}_H = -\hat{P}_\tau \, . \tag{7.54}$$

Using the above formulas, one can calculate the scalar product of spins and isospins in the potential:

$$\hat{V}_{cent} = U_W(r) + U_B(r)\hat{P}_\sigma - U_H(r)\hat{P}_\tau - U_M(r)\hat{P}_\sigma \hat{P}_\tau \, . \tag{7.55}$$

One has still to calculate the r dependencies of the functions: U_W, U_B, U_H, and U_M. The central part of the Gogny force, $U(r)$, dependent on the nucleons' distance, is the Gauss function. Its nonzero range allows to use the full HFB method, i.e., to include the 'pairing' correlations described in Chapter 9. The non-central spin-orbit term is defined in the following way:

$$\hat{V}_{LS} = U_{LS}\, \hat{\mathbf{l}} \cdot (\hat{\mathbf{s}}_1 + \hat{\mathbf{s}}_2), \tag{7.56}$$

where

$$\hat{\mathbf{l}} = \mathbf{r} \times \hat{\mathbf{p}}, \tag{7.57}$$

$\mathbf{r} = \mathbf{r}_1 - \mathbf{r}_2$ is the vector which joins both nucleons and

$$\hat{\mathbf{p}} = \frac{-i\hbar}{2}(\nabla_1 - \nabla_2) \tag{7.58}$$

is their relative momentum p.

The spin-orbit potential U_{LS} can be also an arbitrary scalar function of: r, p, l but usually is taken as $U_{LS} = W_{LS}\delta(\mathbf{r})$, where $\delta(\mathbf{r})$ is the zero-range Dirac delta function.

In Gogny force appears the term dependent on the local density ρ inspired by the Brueckner theory, giving the effective three nucleon forces obtained from the pure nucleon-nucleon interaction. The radial dependence of this term is taken in the form of the Dirac δ function to simplify the calculations (confere also Eq. 7.66).

The last term in Eq. (7.40) originates from the electromagnetic interaction between protons. Isospin Pauli matrix τ_z has the eigenstates $+1$ in neutron state and −1 in proton one, which ensures the vanishing of this term when at least one neutron state appears in nucleon-nucleon pair.

The values of the Gogny parameters, called D1S, are:

$$
\begin{aligned}
W_1 &= -1720.30 \ \text{MeV} & W_2 &= 103.639 \ \text{MeV} \\
B_1 &= 1300.00 \ \text{MeV} & B_2 &= -163.483 \ \text{MeV} \\
H_1 &= -1813.53 \ \text{MeV} & H_2 &= 162.812 \ \text{MeV} \\
M_1 &= 1397.60 \ \text{MeV} & M_2 &= -223.934 \ \text{MeV} \\
\mu_1 &= 0.7 \ \text{fm} & \mu_2 &= 1.2 \ \text{fm} \\
t_0 &= 1390.6 \ \text{MeV}\,\text{fm}^{3(1+\gamma)} & x_0 &= 1 \\
\gamma &= 1/3 & W_{LS} &= 130\,\text{MeV}\,\text{fm}^5
\end{aligned}
\tag{7.59}
$$

Looking at these parameters of the Gogny forces, one can see that the widths of Gauss function μ_1 and μ_2 ensure that the first central term simulates short and the second average range forces. For the given two nucleon state, if one of the components containing the same exchange operator is attractive, the second one should be repulsive because of the opposite signs of the parameters: e.g., W_1 and W_2 or H_1 and H_2. x_0 decides that the density-dependent term is nonzero only in the spin symmetric states. When pairing correlations are neglected, the HFB approach reduces to the Hartree-Fock (HF) method.

7.4 SKYRME FORCES

The Skyrme Hartree Fock model, which also belongs to the non-relativistic self-consisting method, is commonly used despite its doubtful validity for nuclei far from stability and its over 300 parameter sets fitted to various features of nuclei.

The matrix element $\langle l'm'|f(\mathbf{r}_1-\mathbf{r}_2)|lm\rangle$ can be expanded in moments thanks to the fast variation of the effective two-body interaction potential with nuclear radius

$$
\begin{aligned}
\langle l'm'|\, f(r)\,|lm\rangle &= \int \varphi^*_{l'}(\mathbf{r}_1)\varphi^*_{m'}(\mathbf{r}_2) f(r)\varphi_l(\mathbf{r}_1)\varphi_m(\mathbf{r}_2) d^3r\, d^3R \\
&\approx \int f(r)\left[1+(\mathbf{r}\nabla_\mathbf{r})+\tfrac{1}{2}(\mathbf{r}\nabla_\mathbf{r})^2\right]\left[\varphi^*_{l'}(\mathbf{r}_1)\varphi^*_{m'}(\mathbf{r}_2)\right]_{\mathbf{r}=0} \\
&\quad\times\left[1+\mathbf{r}\nabla_\mathbf{r}+\tfrac{1}{2}(\mathbf{r}\nabla_\mathbf{r})^2\right]\left[\varphi_l(\mathbf{r}_1)\varphi_m(\mathbf{r}_2)\right]_{\mathbf{r}=0} d^3r\, d^3R,
\end{aligned}
\tag{7.60}
$$

where $\mathbf{r}=\mathbf{r}_1-\mathbf{r}_2$ and $\mathbf{R}=\frac{1}{2}(\mathbf{r}_1+\mathbf{r}_2)$.

Since for fixed **R**

$$\begin{aligned} \nabla_{\mathbf{r}}[\varphi_l(\mathbf{r}_1)\varphi_m(\mathbf{r}_2)] &= \frac{\partial\varphi_l(\mathbf{r}_1)\varphi_m(\mathbf{r}_2)}{\partial\mathbf{r}_1}\frac{\partial\mathbf{r}_1}{\partial\mathbf{r}} + \frac{\partial\varphi_l(\mathbf{r}_1)\varphi_m(\mathbf{r}_2)}{\partial\mathbf{r}_2}\frac{\partial\mathbf{r}_2}{\partial\mathbf{r}} \\ &= \frac{1}{2}[\partial_{\mathbf{r}_1} - \partial_{\mathbf{r}_2}][\varphi_l(\mathbf{r}_1)\varphi_m(\mathbf{r}_2)], \end{aligned} \tag{7.61}$$

one defines operators $\overleftarrow{\nabla}_{12}=\overleftarrow{\nabla}_1 - \overleftarrow{\nabla}_2$ and $\overrightarrow{\nabla}_{12}=\overrightarrow{\nabla}_1 - \overrightarrow{\nabla}_2$ acting on the bra-states $\varphi^*_{l'}(\mathbf{r}_1)\varphi^*_{m'}(\mathbf{r}_2)$ and on the ket-states $\varphi_l(\mathbf{r}_1)\varphi_m(\mathbf{r}_2)$, respectively. Performing the integration with respect to **r** first and calling the first non-vanishing moments of the function $f(r)$

$$V_0 = \int f(r)d^3r, \qquad \text{and} \qquad 2V_2 = \int r^2 f(r)d^3r$$

the integral can be rewritten as

$$\begin{aligned} &\int d^3R\left[\varphi^*_{l'}(\mathbf{r}_1)\varphi^*_{m'}(\mathbf{r}_2)\right]_{\mathbf{r}=0}\left[V_0 + \frac{V_2}{4}(\overleftarrow{\nabla}^2 + \overrightarrow{\nabla}^2 + 2\overleftarrow{\nabla}\overrightarrow{\nabla})\right][\varphi_l(\mathbf{r}_1)\varphi_m(\mathbf{r}_2)]_{\mathbf{r}=0} \\ = &\int d^3r_1 d^3r_2 \delta(\mathbf{r}_1 - \mathbf{r}_2)\left[\varphi^*_{l'}(\mathbf{r}_1)\varphi^*_{m'}(\mathbf{r}_2)\right]\left[V_0 + \frac{V_2}{4}(\overleftarrow{\nabla}^2 + \overrightarrow{\nabla}^2 + 2\overleftarrow{\nabla}\overrightarrow{\nabla})\right] \\ &\times [\varphi_l(\mathbf{r}_1)\varphi_m(\mathbf{r}_2)]. \end{aligned} \tag{7.62}$$

All odd moments of $f(r)$ vanish. The action of the exchange operator $\hat{P}$ simplifies considerably for zero-range potential-functions f. Let us introduce the spatial ($\hat{P}_r$), spin $\hat{P}_\sigma$), and isospin ($\hat{P}_\tau$) exchange operators, which fullfill

$$\hat{P}_r^2 = \hat{P}_\sigma^2 = \hat{P}_\tau^2 = 1\,.$$

The product of these three operators, applied to an antisymmetric two-nucleon state, gives

$$\hat{P}_r\hat{P}_\sigma\hat{P}_\tau = -1\,.$$

From these relations, follows:

$$\begin{aligned} &\hat{P}_r\delta(\mathbf{r}_1 - \mathbf{r}_2) = \delta(\mathbf{r}_1 - \mathbf{r}_2), &&\hat{P}_r\overleftarrow{\nabla}^2\delta(\mathbf{r}_1 - \mathbf{r}_2) = \overleftarrow{\nabla}^2\delta(\mathbf{r}_1 - \mathbf{r}_2), \\ &\hat{P}_r\overleftarrow{\nabla}\delta(\mathbf{r}_1 - \mathbf{r}_2) = -\overleftarrow{\nabla}\delta(\mathbf{r}_1 - \mathbf{r}_2), &&\hat{P}_\tau\overleftarrow{\nabla}\delta(\mathbf{r}_1 - \mathbf{r}_2) = \hat{P}_\sigma\overleftarrow{\nabla}\delta(\mathbf{r}_1 - \mathbf{r}_2), \\ &\hat{P}_\tau\delta(\mathbf{r}_1 - \mathbf{r}_2) = -\hat{P}_\sigma\delta(\mathbf{r}_1 - \mathbf{r}_2), &&\hat{P}_\tau\overleftarrow{\nabla}^2\delta(\mathbf{r}_1 - \mathbf{r}_2) = -\hat{P}_\sigma\overleftarrow{\nabla}^2\delta(\mathbf{r}_1 - \mathbf{r}_2), \end{aligned} \tag{7.63}$$

and the same equations with $\overrightarrow{\nabla}$.

The general form of the two-particle effective potential comes from these results, together with Eq. (7.61), in the approximation of very short range,

$$\begin{aligned} V_{12} &= t_0(1+x_0\hat{P}_\sigma)\delta(\vec{r}_1-\vec{r}_2) \\ &- \frac{1}{8}t_1(1+x_1\hat{P}_\sigma)\left[\overleftarrow{\nabla}_{12}^2\,\delta(\vec{r}_1-\vec{r}_2)+\delta(\vec{r}_1-\vec{r}_2)\,\overrightarrow{\nabla}_{12}^2\right] \\ &+ \frac{1}{4}t_2(1+x_2\hat{P}_\sigma)\,\overleftarrow{\nabla}_{12}\,\delta(\vec{r}_1-\vec{r}_2)\,\overrightarrow{\nabla}_{12} \end{aligned} \tag{7.64}$$

with

$$\begin{aligned} t_0 &= (W+M)V_0, & t_0x_0 &= (B+H)V_0 \\ -t_1 &= 2(W+M)V_2, & -t_1x_1 &= 2(B+H)V_2 \\ t_2 &= 2(W-M)V_2, & t_2x_2 &= 2(B-H)V_2\,. \end{aligned}$$

Bell and Skyrme added a spin-orbit term to control the size of the spin-orbit splitting in the HF single-particle potential, derived from the two-body interaction (2.105),

$$V_{LS} = i(1/4)W_{LS}\,[\overleftarrow{\vec{\nabla}}_{12}\times\delta(\vec{r}_1-\vec{r}_2)\,\overrightarrow{\vec{\nabla}}_{12}]\cdot(\vec{\sigma}_1+\vec{\sigma}_2) \tag{7.65}$$

to the central potential (2.105), where $\vec{\sigma}_i$ is twice the the spin operator of nucleon i. Skyrme also added a repulsive three-body force of zero range $V_{123} = t_3\delta(\mathbf{r}_1-\mathbf{r}_2)\delta(\mathbf{r}_2-\mathbf{r}_3)$ to the two-body interaction to enforce proper saturation density and energy density in bulk nuclear matter. This term was shown by Baeckmann et al. for spin-saturated systems to be equivalent to a density-dependent two-particle potential

$$V_{123} = \frac{t_3}{6}(1+x_3\hat{P}_\sigma)\rho(\mathbf{R})\delta(\mathbf{r}_1-\mathbf{r}_2), \tag{7.66}$$

where $\rho(\mathbf{R})$ is the sum of proton and neutron densities. One generally uses the slightly more general form to obtain more flexibility to reproduce the position of giant monopole resonance energies,

$$V_{123} = \frac{t_3}{6}(1+x_3\hat{P}_\sigma)[\rho(\mathbf{R})]^\gamma\delta(\mathbf{r}_1-\mathbf{r}_2) \tag{7.67}$$

with additional parameters x_3 and γ.
The Coulomb two-body potential between protons is given by

$$V_{\text{Coul}} = \frac{e^2}{|\mathbf{r}_1-\mathbf{r}_2|}\,. \tag{7.68}$$

The total effective interaction is the sum of the potentials (7.65)-(7.68)

$$V_{\text{Skyrme}} = V_{12} + V_{LS} + V_{123} + V_{\textbf{Coul}} \,. \tag{7.69}$$

The parameters of the Skyrme interaction are: $t_0, t_1, t_2, t_3, x_0, x_1, x_2, x_3, W_{LS}$ and γ. Several sets of these parameters are adjusted to binding energies of nuclei from various mass regions and additional observables, particularly random phase RPA-type excitations of the monopole, dipole, or quadrupole character. Frequently used sets are, e.g., SkIII (fairly good binding energies, less perfect radii), SkM* (good binding energies, including fission barriers, good radii), SkP (gives realistic pairing with Skyrme interactions), or SLy6 (fitted to describe neutron-rich nuclei and neutron stars).

For fission studies, the SkM* potential has often been used. Its set of parameters

$$\begin{array}{ll} t_0 = \text{-2645 MeV fm}^3 & x_0 = 0.09 \\ t_1 = \text{410 MeV fm}^5 & x_1 = 0 \\ t_2 = \text{-135 MeV fm}^5 & x_2 = 0 \\ t_3 = \text{15595 MeV fm}^{(3+3\gamma)} & x_3 = 0 \\ W_{LS} = \text{130 MeV fm}^5 & \gamma = 1/6 \end{array}$$

was chosen to reproduce in particular experimental fission barrier heights.

The Slater determinants in calculating the expectation value of the Hamiltonian with Skyrme interactions lead to the following energy density functional (EDF)

$$E = \int \mathcal{E}[\rho(\mathbf{r}), \nabla\rho(\mathbf{r}), \tau(\mathbf{r}), \mathbf{J}(\mathbf{r})]\, d^3r \,, \tag{7.70}$$

where the energy density $\mathcal{E}$ is a function of the nucleon densities ρ_q, the kinetic energy densities τ_q, and spin-orbit densities

$$\mathbf{J}_q = -i \sum_{i,\sigma,\sigma'} \varphi_i(\mathbf{r},\sigma,q)[\vec{\nabla}\varphi_i(\mathbf{r},\sigma',q) \times \langle\sigma|\sigma|\sigma'\rangle] \,. \tag{7.71}$$

The explicit form of $\mathcal{E}$ was derived by Vautherin and Brink assuming that there is no isospin mixing in the HF basis and that the ground state is time-reversal invariant

$$\begin{aligned} &\mathcal{E}[\rho_q(\vec{r}), \tau_q(\vec{r}), \vec{J}_q(\vec{r})] = \\ &\quad \frac{\hbar^2}{2M_{\text{nucl}}}\tau + \frac{1}{2}t_0\left[\left(1+\frac{x_0}{2}\right)\rho^2 - \left(x_0+\frac{1}{2}\right)(\rho_n^2+\rho_p^2)\right] \\ &\quad + \frac{1}{12}t_3\rho^\gamma\left[\left(1+\frac{x_3}{2}\right)\rho^2 - \left(x_3+\frac{1}{2}\right)(\rho_n^2+\rho_p^2)\right] \end{aligned}$$

$$
\begin{aligned}
&+\frac{1}{4}\left[t_1\left(1+\frac{x_1}{2}\right)+t_2\left(1+\frac{x_2}{2}\right)\right]\tau\rho \\
&-\frac{1}{4}\left[t_1\left(x_1+\frac{1}{2}\right)-t_2\left(x_2+\frac{1}{2}\right)\right](\tau_n\rho_n+\tau_p\rho_p) \\
&+\frac{1}{16}\left[3t_1\left(1+\frac{x_1}{2}\right)-t_2\left(1+\frac{x_2}{2}\right)\right](\vec{\nabla}\rho)^2 \\
&-\frac{1}{16}\left[3t_1\left(x_1+\frac{1}{2}\right)+t_2\left(x_2+\frac{1}{2}\right)\right][(\vec{\nabla}\rho_n)^2+(\vec{\nabla}\rho_p)^2] \\
&+\frac{W_{LS}}{2}[\vec{J}\cdot\vec{\nabla}\rho+\vec{J}_n\cdot\vec{\nabla}\rho_n+\vec{J}_p\cdot\vec{\nabla}\rho_p]+\mathcal{E}_{\mathrm{Coul}}\,,
\end{aligned}
\tag{7.72}
$$

where $\rho = \rho_p + \rho_n$, $\tau = \tau_p + \tau_n$, and $\mathcal{E}_{\mathrm{Coul}}$ is the Coulomb energy density. It can be written as the sum of direct and exchange contributions. The exchange term is usually approximated by a form obtained for the Slater determinants

$$
\mathcal{E}_{\mathrm{Coul}}(\vec{r}) = \frac{e^2}{2}\rho_p(\vec{r})\int d^3r'\,\frac{\rho_p(\vec{r}\,')}{|\vec{r}-\vec{r}\,'|} - \frac{3}{4}e^2\left(\frac{3}{\pi}\right)^{1/3}\rho_p^{4/3}(\vec{r})\,. \tag{7.73}
$$

For odd systems whose ground-state wave function is not time-reversal invariant, the energy density has additional terms. In the first two lines on the right-hand-side of Eq. (7.72) one determines properties of bulk nuclear matter, the saturation density ρ_0, the volume-energy coefficient of the liquid-drop formula c_1, and the incompressibility parameter K. The terms in the third and fourth lines control the effective mass of nucleon $M^*_{\mathbf{nucl}}$, and those in the fifth and sixth lines contribute to the surface energy. One obtains the Skyrme HF-equation minimizing the total energy of Eq. (7.70) with respect to variations of the single-particle wave-functions $\varphi_j^{(q)}$,

$$
\hat{h}_q\varphi_j^{(q)} = \varepsilon_j^{(q)}\varphi_j^{(q)}\,, \tag{7.74}
$$

for the mean-field single-particle Hamiltonian

$$
\hat{h}_q = -\vec{\nabla}\frac{\hbar^2}{2M^{(q)*}_{\mathrm{nucl}}(\vec{r})}\vec{\nabla} + V_q(\vec{r}) - i\vec{W}_q(\vec{r})\cdot(\vec{\nabla}\times\vec{\sigma})\,. \tag{7.75}
$$

Functional derivative of the energy (7.70) with respect to the density gives the central nuclear potential $V_q(\mathbf{r})$

$$
V_q(\vec{r}) = \frac{\delta E}{\delta\rho_q(\vec{r})} = \frac{\partial\mathcal{E}}{\partial\rho_q} - \vec{\nabla}\left(\frac{\partial\mathcal{E}}{\partial(\vec{\nabla}\rho_q)}\right). \tag{7.76}
$$

For Skyrme potentials this yields

$$
\begin{aligned}
V_q(\vec{r}) &= t_0\left(1+\frac{x_0}{2}\right)\rho - t_0\left(x_0+\frac{1}{2}\right)\rho_q \\
&+ \frac{t_3}{12}\gamma\rho^{\gamma-1}\left[\left(1+\frac{x_3}{2}\right)\rho^2 - \left(x_3+\frac{1}{2}\right)(\rho_n^2+\rho_p^2)\right] \\
&+ \frac{t_3}{6}\rho^{\gamma}\left[\left(1+\frac{x_3}{2}\right)\rho - \left(x_3+\frac{1}{2}\right)\rho_q\right] \\
&+ \frac{1}{4}\left[t_1\left(1+\frac{x_1}{2}\right)+t_2\left(1+\frac{x_2}{2}\right)\right]\tau \\
&- \frac{1}{4}\left[t_1\left(x_1-\frac{1}{2}\right)+t_2\left(x_2+\frac{1}{2}\right)\right]\tau_q \\
&- \frac{1}{8}\left[3t_1\left(1+\frac{x_1}{2}\right)-t_2\left(1+\frac{x_2}{2}\right)\right](\vec{\nabla}^2\rho) \\
&+ \frac{1}{8}\left[3t_1\left(x_1+\frac{1}{2}\right)+t_2\left(x_2+\frac{1}{2}\right)\right](\vec{\nabla}^2\rho_q) \\
&- \frac{W_{LS}}{2}[\vec{\nabla}\cdot\vec{J}+\vec{\nabla}\cdot\vec{J}_q] + V_{\text{Coul}}\,,
\end{aligned}
\tag{7.77}
$$

where the Coulomb potential is given by

$$
V_{\text{Coul}}(\vec{r}) = e^2\int\frac{\rho_p(\vec{r}\,')}{|\vec{r}-\vec{r}\,'|}d^3r' - e^2\left(\frac{3}{\pi}\right)^{1/3}\rho_p^{1/3}(\vec{r})\,. \tag{7.78}
$$

The effective-mass form factor is defined as

$$
\begin{aligned}
f_q(\vec{r}) &= \frac{M_{\text{nucl}}^{(q)}}{M_{\text{nucl}}^{(q)*}(\vec{r})} = \frac{2M_{\text{nucl}}^{(q)}}{\hbar^2}\frac{\delta E}{\delta\tau_q(\vec{r})} \\
&= 1+\frac{2M_{\text{nucl}}^{(q)}}{\hbar^2}\left\{\frac{1}{4}\left[t_1(1+\frac{x_1}{2})+t_2(1+\frac{x_2}{2})\right]\rho(\vec{r})\right. \\
&\quad\left.-\frac{1}{4}\left[t_1(x_1+\frac{1}{2})-t_2(x_2+\frac{1}{2})\right]\rho_q(\vec{r})\right\}
\end{aligned}
\tag{7.79}
$$

and the spin-orbit potential as

$$
\vec{W}_q(\vec{r}) = \frac{\delta E}{\delta\vec{J}_q(\vec{r})} = \frac{1}{2}W_{LS}\;\vec{\nabla}(\rho+\rho_q)\,. \tag{7.80}
$$

Summing up, the mean-field theory of the Skyrme type starts from an energy functional, which is derived from a density-dependent two-body interaction of the form:

$$
\begin{aligned}
V(i,j) &= t_0(1+x_0\hat{P}_\sigma)\delta(\vec{r}_i-\vec{r}_j) \\
&- \frac{1}{2}t_1(1+x_1\hat{P}_\sigma)\left[\overleftarrow{\nabla}_{ij}^2\delta(\vec{r}_i-\vec{r}_j)+\delta(\vec{r}_i-\vec{r}_j)\overrightarrow{\nabla}_{ij}^2\right] \\
&- t_2(1+x_2\hat{P}_\sigma)\overleftarrow{\nabla}_{ij}\delta(\vec{r}_i-\vec{r}_j)\overrightarrow{\nabla}_{ij} \qquad (7.81) \\
&+ \frac{1}{6}t_3P_\sigma)\left[\rho_{q_i}(\vec{r}_i)+\rho_{q_j}(\vec{r}_j)\right]^\gamma\delta(\vec{r}_i-\vec{r}_j) \\
&- i\omega_0\overleftarrow{\nabla}_{ij}\wedge\delta(\vec{r}_i-\vec{r}_j)\overrightarrow{\nabla}_{ij}\cdot(\vec{\sigma}_i+\vec{\sigma}_j)+V_{\mathrm{Coul}}\,,
\end{aligned}
$$

where $t_0, t_1, t_2, t_3, x_1, x_2, x_3, w_0$, and γ are the adjustable parameters, and P_σ is the spin exchange operator.

The total energy (7.70), expressed by the energy density functional of a nucleus (7.72), has to be minimized with respect to the choice of the many-body wave functions, which leads to a non-linear eigenproblem with the corresponding mean-field Hamiltonian.

7.5 RELATIVISTIC MEAN FIELD THEORY

This section demands from the reader some basic knowledge of elementary particle physics, quantum field theory, relativistic Dirac and Klein-Gordon equations, and the Hartree-Fock method.

7.5.1 Nucleon-nucleon interaction

The attractive part of nucleon-nucleon interaction, shown in Fig. 1.1a origins from the π-meson field. These mesons are the exchange particles, enabling the interaction between nucleons. The relativistic mean-field theory (RMFT) is based on a model consisting of the nucleons (ψ), electromagnetic (A^μ) fields, and also the $\vec{\pi}$, ω and $\vec{\rho}$ meson fields. It reproduces well the main features of nuclear forces. The pions π describe the long-range interactions while the ω field rules the spin-orbit and repulsive terms. $\vec{\rho}$ field is responsible for the isospin effects. In the RMFT model, the $\vec{\pi}$-meson fields are not taken directly into account, but effective σ mesons replace them.

7.5.2 Lagrange function

The RMFT is based on relativistic quantum mechanics. A unique field represents each type of particle, and an appropriate Lagrangian density ($\mathcal{L}$) is defined for each one. The equations of motion based on these Lagrangians describe the features of the field sources. In the RMFT, the Lagrangian density is a sum of terms corresponding to the different types of particles, the term responsible for the interaction between the nucleon and meson, and the part originating from the electromagnetic field:

$$\mathcal{L} = \mathcal{L}_{\mathrm{n}} + \mathcal{L}_{\sigma} + \mathcal{L}_{\omega} + \mathcal{L}_{\rho} + \mathcal{L}_{\mathrm{int}} + \mathcal{L}_{\mathrm{e.m.}} \,. \tag{7.82}$$

The nucleons are represented by the baryons fields, with eigenfunctions being the single-particle Dirac spinors:

$$\psi_k(x) = (\psi_k^1(x), \psi_k^2(x), \psi_k^3(x), \psi_k^4(x)) \,. \tag{7.83}$$

Here k stays for quantum numbers of the nucleon state. The Lagrangian term representing the nucleon field is:

$$\mathcal{L}_{\mathrm{n}} = \sum_{k=1}^{A} \overline{\psi}_k \{i\gamma^\mu \partial_\mu - m\} \psi_k \,. \tag{7.84}$$

The summation goes over all occupied nucleon levels, and m is the nucleon's mass. The repeating indexes $\mu = 0, 1, 2, 3$ denote the summation.
The Dirac matrices (γ^μ) have four components with 4×4 dimensions:

$$\gamma^0 = \begin{pmatrix} I & \oslash \\ \oslash & -I \end{pmatrix}, \gamma^n = \begin{pmatrix} \oslash & \tau^n \\ -\tau^n & \oslash \end{pmatrix}, \tag{7.85}$$

where I is an identity matrix, $\oslash$ is a 2×2 zeros matrix, and τ^n are the Pauli matrices.

$$I = \begin{pmatrix} 1 & 0 \\ 0 & 1 \end{pmatrix}, \; \tau^1 = \begin{pmatrix} 0 & 1 \\ 1 & 0 \end{pmatrix}, \; \tau^2 = \begin{pmatrix} 0 & -i \\ i & 0 \end{pmatrix}, \; \tau^3 = \begin{pmatrix} 1 & 0 \\ 0 & -1 \end{pmatrix}. \tag{7.86}$$

The function $\overline{\psi}_k(x)$ is the Dirac coupling of the $\psi_k(x)$ function:

$$\overline{\psi}_k(x) = \psi_k^+(x) \gamma^0, \tag{7.87}$$

and ∂_μ are the partial derivatives over space-time coordinates $(x^\mu) = (t, x, y, z)$

$$\partial_\mu = \frac{\partial}{\partial x^\mu}. \tag{7.88}$$

The Lagrangian density terms originating from exchange particles σ, ω, ρ fields are taken in the RMFT in the following way:

$$\mathcal{L}_\sigma = \frac{1}{2}\partial^\mu\sigma\partial_\mu\sigma - U(\sigma), \tag{7.89}$$

$$\mathcal{L}_\omega = -\left(\frac{1}{2}\Omega^{\mu\nu}\Omega_{\mu\nu} - \frac{1}{4}m_\omega^2\omega^\mu\omega_\mu\right), \tag{7.90}$$

$$\mathcal{L}_\rho = -\left(\frac{1}{2}\vec{R}^{\mu\nu}\vec{R}_{\mu\nu} - \frac{1}{4}m_\rho^2\vec{\rho}^\mu\vec{\rho}_\mu\right). \tag{7.91}$$

The part of the Lagrangian responsible for the interaction between the meson fields and nucleons is usually assumed in a simple linear form:

$$\mathcal{L}_{\text{int}} = -\bar{\psi}(g_\sigma\sigma + g_\omega\gamma_\mu\omega^\mu + g_\rho\gamma_\mu\vec{\tau}\vec{\rho}^\mu + e\gamma_\mu A^\mu)\psi\,. \tag{7.92}$$

The Coulomb term describing protons' electromagnetic interaction contains the mass-less photons

$$\mathcal{L}_{\text{e.m.}} = -\frac{1}{2}F^{\mu\nu}F_{\mu\nu}\,. \tag{7.93}$$

The scalar field potential $U(\sigma)$ in Eq. (7.89) is usually assumed in the form:

$$U(\sigma) = \frac{1}{2}m_\sigma^2\sigma^2 + \frac{1}{3}g_2\sigma^3 + \frac{1}{4}g_3\sigma^4\,. \tag{7.94}$$

The RMFT Lagrangian equations contain nine parameters, but only five of them are adjustable. They are: masses of particles m_σ, m_ω, m_ρ and potential parameters g_2 and g_3. The fields ω^μ, $\vec{\rho}^\mu$ and the electromagnetic potential A^μ are described by the following second-range tensors:

$$\Omega^{\mu\nu} = \partial^\mu\omega^\nu - \partial^\nu\omega^\mu\,, \tag{7.95}$$

$$\vec{R}^{\mu\nu} = \partial^\mu\vec{\rho}^\nu - \partial^\nu\vec{\rho}^\mu - g_\rho(\vec{\rho}^\mu \times \vec{\rho}^\nu)\,, \tag{7.96}$$

$$F^{\mu\nu} = \partial^\mu A^\nu - \partial^\nu A^\mu\,, \tag{7.97}$$

where g_ρ is the coupling constant of ρ field.

The interaction of exchange particles with nucleons demands additional terms in the Lagrangian, which in the linear approximation have the following form:

$$\begin{aligned} \mathcal{L}_{int} &= -g_\sigma\overline{\psi}_k\psi_k\sigma \\ &- g_\omega\overline{\psi}_k\gamma^\mu\psi_k\omega_\mu \\ &- g_\rho\overline{\psi}_k\gamma_\mu\vec{\tau}\psi_k\vec{\rho}_\mu \\ &- e\overline{\psi}_k\gamma_\mu\frac{(1-\tau_3)}{2}\psi_k A_\mu\,. \end{aligned} \tag{7.98}$$

The bilinear forms of field ψ ($\overline{\psi}_k\psi_k$, $\overline{\psi}_k\gamma^\mu\psi_k$) and ($\overline{\psi}_k\gamma^\mu\vec{\tau}\psi_k$) correspond to the interaction of the spinor field with the scalar σ, vector ω_μ and isovector $\vec{\rho}_\mu$ fields. The interactions between exchange particles are small and not taken into account in calculations.

7.5.3 Fields equations and their solution methods

Knowing the Lagrangian density, one can find the motion equations for nucleons and exchange particles using the least-action Hamilton principle

$$\delta\mathcal{S} = 0\,, \tag{7.99}$$

where the action $\mathcal{S}$ is expressed by the Lagrangian density $\mathcal{L}$ integral

$$\mathcal{S} = \int_\Omega d^4x\,\mathcal{L}\big(\phi(x), \partial_\mu\phi(x), t\big)\,, \tag{7.100}$$

where $\phi_j(x)$ and $\partial_\mu\phi_j(x)$ are the generalized coordinates and momenta corresponding to different particle fields. The integration goes over the 4-dimensional time-space volume Ω in which the particles move.
Having the Lagrange function, one can evaluate in a standard way the energy-momentum tensor

$$T^{\mu\nu} = -g^{\mu\nu}\mathcal{L} + \frac{\partial\mathcal{L}}{\partial(\partial_\mu\phi_j)}\partial^\mu\phi_j\,. \tag{7.101}$$

The above energy-momentum tensor obeys the continuity equation $\partial_\mu T^{\mu\nu} = 0$ and allows for evaluations of the four-momentum

$$P^\nu = \int d^3r\,T^{0\nu}(\mathbf{r})\,, \tag{7.102}$$

whose 0-th component is the energy of the nucleus

$$E = P^0 = \int d^3r\,\mathcal{H}(\mathbf{r})\,, \tag{7.103}$$

where $\mathcal{H}$ is the energy density functional

$$\mathcal{H} = T^{00} = \frac{\partial\mathcal{L}}{\partial\dot{\phi}_j}\dot{\phi}_j - \mathcal{L}\,. \tag{7.104}$$

The Hamilton principle (7.99) leads to the Euler-Lagrange motion equations for particular fields corresponding to different kinds of particles:

$$\partial_\mu\left(\frac{\partial\mathcal{L}}{\partial(\partial_\mu\phi_j(x))}\right) - \frac{\partial\mathcal{L}}{\partial\phi_j(x)} = 0 \quad \text{for all } j\,. \tag{7.105}$$

After inserting the Lagrangian density (7.82) into the above equation, one derives the Euler-Lagrange type equation of motion for every particular field: nucleons ψ_k, scalar σ, vector ω_μ, and isovector $\vec{\rho}_\mu$ mesons, and electromagnetic massless vector A_μ. In the ground state, which is time-invariant, one can get static solutions of these motion equations.

For nucleons, one gets the Dirac equation:

$$\{\gamma^0(-i\boldsymbol{\nabla} - \boldsymbol{V}(\boldsymbol{r})) + \gamma m^*(\boldsymbol{r}) + S(\boldsymbol{r})\}\psi_k(\boldsymbol{r}) = \varepsilon_k \psi_k(\boldsymbol{r}) , \tag{7.106}$$

where the scalar potential S depends on the time components of fields:

$$S(\boldsymbol{r}) = g_\omega \omega_0(\boldsymbol{r}) + g_\rho \vec{\tau}\vec{\rho}_0(\boldsymbol{r}) + eA_0(\boldsymbol{r})\frac{1-\tau_3}{2} , \tag{7.107}$$

while the vector potential $\boldsymbol{V}(\boldsymbol{r})$ is expressed by the space components:

$$\boldsymbol{V}(\boldsymbol{r}) = g_\omega \boldsymbol{\omega}(\boldsymbol{r}) + g_\rho \vec{\tau}\vec{\boldsymbol{\rho}}(\boldsymbol{r}) + e\boldsymbol{A}(\boldsymbol{r})\frac{1-\tau_3}{2} . \tag{7.108}$$

The constant e in the above two equations represents the elementary charge. The effective mass of nucleon depends on σ scalar field:

$$m^*(\boldsymbol{r}) = m + g_\sigma \sigma(\boldsymbol{r}) . \tag{7.109}$$

The meson fields are described by the Klein-Gordon equations:

$$(-\Delta + m_\sigma^2)\sigma(\boldsymbol{r}) = -g_\sigma \rho_\sigma(\boldsymbol{r}) - g_2\sigma^2(\boldsymbol{r}) - g_3\sigma^3(\boldsymbol{r}) , \tag{7.110}$$

$$(-\Delta + m_\omega^2)\omega_\mu(\boldsymbol{r}) = g_\omega j^\mu(\boldsymbol{r}) , \tag{7.111}$$

$$(-\Delta + m_\rho^2)\boldsymbol{\rho}_\mu(\boldsymbol{r}) = g_\rho \vec{j}^\mu(\boldsymbol{r}) . \tag{7.112}$$

The Maxwell equations describe the evolution of the electromagnetic field:

$$-\Delta A^\mu(\boldsymbol{r}) = e j_p^\mu(\boldsymbol{r}). \tag{7.113}$$

The source terms in the left-hand side of the four above equations depend on the nucleon wave functions ψ_i. They are:
- scalar field density

$$\rho_\sigma(\boldsymbol{r}) = \sum_{i=1}^{A} \overline{\psi}_i(\boldsymbol{r})\psi_i(\boldsymbol{r}) , \tag{7.114}$$

- four-vector

$$j^\mu(\boldsymbol{r}) = \sum_{i=1}^{A} \overline{\psi}_i(\boldsymbol{r})\gamma^\mu \psi_i(\boldsymbol{r}) , \tag{7.115}$$

- four-isovector

$$\vec{j}^{\mu}(\boldsymbol{r}) = \sum_{i=1}^{A} \overline{\psi}_i(\boldsymbol{r}) \gamma^{\mu} \vec{\tau} \psi_i(\boldsymbol{r}) , \tag{7.116}$$

- and proton current four-vector :

$$j_p^{\mu}(\boldsymbol{r}) = \sum_{i=1}^{A} \overline{\psi}_i(\boldsymbol{r}) \gamma^{\mu} \frac{1-\tau_3}{2} \psi_i(\boldsymbol{r}) . \tag{7.117}$$

The above equations do not contain yet the mean-field assumption, whose approximation enables a much easier solution of a nonlinear differential equation system. The meson and Coulomb fields change rather slowly while protons and neutrons move in a mean-field, so their functions can be averaged, omitting the meson field fluctuations. The time-reversal symmetry demands the disappearance of the space components of fields ω, $\vec{\rho}$, and $\mathbf{A}$. The charge conservation keeps only the third component of the isovector field ρ_0.
All these assumptions allow to replace the four-currants by their time-components only:

$$\begin{aligned} \rho_\omega(\boldsymbol{r}) &= \sum_{i=1}^{A} \overline{\psi}_i(\boldsymbol{r}) \psi_i(\boldsymbol{r}) , \\ \rho_\rho(\boldsymbol{r}) &= \sum_{i=1}^{A} \overline{\psi}_i(\boldsymbol{r}) \tau_3 \psi_i(\boldsymbol{r}) , \\ \rho_p(\boldsymbol{r}) &= \sum_{i=1}^{A} \overline{\psi}_i(\boldsymbol{r}) \tfrac{1-\tau_3}{2} \psi_i(\boldsymbol{r}) . \end{aligned} \tag{7.118}$$

The equations of motion can now be written in the following form:

$$\begin{aligned} &\{-i\boldsymbol{\gamma}\boldsymbol{\nabla} + \gamma^0 m^*(\boldsymbol{r}) + S(\boldsymbol{r})\} \psi_k(\boldsymbol{r}) = \varepsilon_i \psi_k(\boldsymbol{r}) , \\ &(-\Delta + m_\sigma^2)\sigma(\boldsymbol{r}) = -g_\sigma \rho_\sigma(\boldsymbol{r}) - g_2 \sigma^2(\boldsymbol{r}) - g_3 \sigma^3(\boldsymbol{r}) , \\ &(-\Delta + m_\omega^2)\omega_0(\boldsymbol{r}) = g_\omega \rho_\omega(\boldsymbol{r}) , \\ &(-\Delta + m_\rho^2)\vec{\rho}_0(\boldsymbol{r}) = g_\rho \vec{\rho}_\rho(\boldsymbol{r}) , \\ &-\Delta A_0(\boldsymbol{r}) = e\rho_p(\boldsymbol{r}) . \end{aligned} \tag{7.119}$$

Note that the vector potential $\boldsymbol{V}(\boldsymbol{r})$ is now not present in the Dirac equation, while the scalar potential takes the following form:

$$S(\boldsymbol{r}) = g_\omega \omega_0(\boldsymbol{r}) + g_\rho \tau_3 \rho_0(\boldsymbol{r}) + eA_0(\boldsymbol{r}) \frac{(1-\tau_3)}{2} . \tag{7.120}$$

The set of motion equations (7.119) can be solved by an iteration method.

The most efficient numerical method of solving the above problem is based on a Hartree-Fock-like approximation:

Table 7.2 Frequently used RMFT parameter sets

Parameter	Units	NL–1	NL–SH	NL–3
M	MeV	938.0	939.0	939.0
m_σ	MeV	492.25	526.059	508.194
m_ω	MeV	795.359	783.00	782.501
m_ρ	MeV	763.000	763.00	763.000
g_σ	-	10.138	10.4444	10.217
g_2	fm^{-1}	-12.172	-6.9099	-10.431
g_3	-	-36.265	-15.8337	-28.885
g_ω	-	13.285	12.945	12.868
g_ρ	-	4.976	4.383	4.474

- One begins with the nucleon wave function evaluated using an average single-particle potential, e.g., harmonic oscillator,
- Then, the scalar potential in the Dirac equation is calculated, and the protons and neutrons wave functions are found,
- Having the nucleons wave functions, one evaluates the densities of meson fields and inserts them into the Klein-Gordon equations, giving the exchange particle fields,
- The fields obtained in such a way are used in the next iteration step,
- The iteration procedure is finished when the assumed convergence quality is achieved.

In such a way, one obtains the wave functions and single-particle energies necessary to calculate the nuclear observables. For example, the total energy of the nucleus can be evaluated using Eq. (7.103).

The nine parameters of RMFT are: the nucleon and the exchange particles masses (m, m_σ, m_ρ,m_ω), coupling constants (ρ_σ, ρ_ρ, ρ_ω), and scalar potential constants (g_2, g_3). Only six of them are really free adjustable parameters as the masses on nucleon, ρ, and ω mesons are close to the experimental values. A few examples of the RMFT parameter sets are listed in Table 7.2. These parameters were adjusted to the global properties of atomic nuclei from different mass regions.

CHAPTER 8

Residual forces in nuclei

Although one assumes that the particles in the nucleus move independently in the average mean-field potential, it turns out that there exists some residual two-body effective interactions that are not included in the average potential. Building a many-body Hamiltonian $\hat{H}$ of the A nucleon system, one can not limit to the sum of single-particle Hamiltonians $\hat{h}(i)$. One must also take into account the residual forces $\hat{V}_{\mathrm{res}}$

$$\hat{H} = \hat{H}_0 + \hat{V}_{\mathrm{res}}\,, \tag{8.1}$$

where

$$\hat{H}_0 = \sum_{i=1}^{A} \hat{h}(i) \tag{8.2}$$

and

$$\hat{V}_{\mathrm{res}} = \sum_{\substack{i=1\\ i\neq j}}^{A} \hat{v}(ij)\,. \tag{8.3}$$

In the second quantization (7.18, 7.25) the Hamiltonian $\hat{H}$ takes the following form:

$$\hat{H}^{\mathrm{II}} = \sum_{\mu\nu} \langle\mu|\hat{h}|\nu\rangle a_\mu^+ a_\nu + \frac{1}{2}\sum_{\mu\nu\gamma\delta} \langle\mu\nu|\hat{v}|\gamma\delta\rangle a_\mu^+ a_\nu^+ a_\delta a_\gamma\,, \tag{8.4}$$

where the functions $|\nu\rangle, \ldots$ are the complete system of single-particle states, which are, e.g., solutions of the $\hat{h}$ eigenproblem

$$\hat{h}|\nu\rangle = e_\nu|\nu\rangle\,. \tag{8.5}$$

The operators of creation (a_ν^+) and annihilation (a_ν) of fermions in the states $|\nu\rangle$ have to fulfill the anti-commutation conditions (7.7), (7.10), and (7.12).

Since the matrix element $\langle\mu\nu|\hat{v}|\gamma\delta\rangle$ can be comparable with single-particle level distances (e.g., $\hbar\omega_0$), one can not use a standard perturbation calculation to find the solution of the eigenproblem of $\hat{H}$. There are basically two most important methods of inclusion of residual forces in the nucleus:

— for single, well-separated shell, an accurate diagonalization can be made, assuming a specific form of forces $\hat{V}_{\text{res}}$,

— by dividing, according to Bohr and Mottelsson, the residual interactions into long-range forces and short-range forces

$$\hat{V}_{\text{res}} = \hat{V}_{\text{long}} + \hat{V}_{\text{short}}\,, \tag{8.6}$$

where the range of $\hat{V}_{\text{short}}$ is much smaller than the radius R of a nucleus and the $\hat{V}_{\text{long}}$ range is R.

Different methods account for both types of residual forces. The energy of the long-range interaction of nucleons is proportional to the square of the number of nucleons outside the core, and the short-range interaction energy is proportional to the number of nucleons inside the core. One will include the long-range interaction using particle-hole coupling and the short-range correlations.

8.1 EFFECT OF THE LONG-RANGE FORCES

The attracting long-range residual forces create a stable deformation of atomic nuclei with nucleon numbers different from magic ones. The mechanism of nucleus deformation formation during the addition of subsequent nucleons is shown in Fig. 9.1. First, outside the closed shell, the nucleon will place itself in the equipotential plane in orbit j, i.e., it will take the state of the third component $m = j$. As a result of long-range attraction forces, subsequent nucleons will try to fill the spatially closest states $m = \pm j, \pm|j-1|, \ldots$. In this way, the density of nuclei will increase the outer equatorial area. The nucleus will become flattened (*oblate*) as shown in Fig. 8.1. A similar situation occurs when one subtracts more nucleons from a doubly magic spherical nucleus. Then, density will decrease in the outer part of the nucleus, and the nucleus will take an elongated shape (*prolate*).

One assumes that $\hat{V}_{\text{long}}$ has a range comparable with the radius R of the nucleus and depends on distances between the nucleons being in the $\mathbf{r}_i$ and $\mathbf{r}_j$ points, i.e., from

$$r_{ij} = |\mathbf{r}_i - \mathbf{r}_j| = \sqrt{r_i^2 + r_j^2 + 2\mathbf{r}_i \cdot \mathbf{r}_j \cos\vartheta_{ij}}\,. \tag{8.7}$$

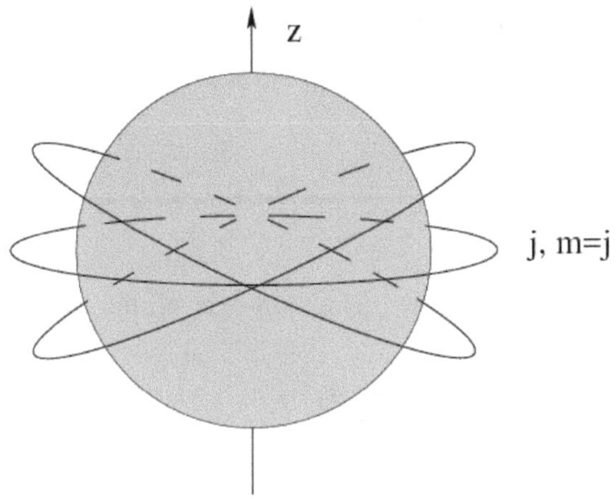

Figure 8.1 Long-range forces effect.

Let us assume that the long-range force between the i-th and the j-th nucleon depends only on the distance of both particles:

$$V_l(r_{ij}) = V_l(|\mathbf{r}_i - \mathbf{r}_j|)\,. \tag{8.8}$$

One can expand the V_l interaction into a series of Legendre polynomials $P_\lambda(\cos\vartheta_{ij})$

$$V_l(|\mathbf{r}_i - \mathbf{r}_j|) = \sum_{\lambda=0}^{\infty} v_\lambda(r_i, r_j)\, P_\lambda(\cos\vartheta_{ij})\,. \tag{8.9}$$

This series is fast converging. Each expansion term can be presented in the form of product of spherical functions (2.78)

$$P_\lambda(\cos\vartheta_{ij}) = \frac{4\pi}{2\lambda+1} \sum_{\mu=-\lambda}^{\lambda} Y^\star_{\lambda\mu}(\vartheta_i, \varphi_i) Y_{\lambda\mu}(\vartheta_j, \varphi_j)\,, \tag{8.10}$$

hence, one gets

$$V_l(|\mathbf{r}_i - \mathbf{r}_j|) = \sum_{\lambda=0}^{\infty} v_\lambda(r_i, r_j) \frac{4\pi}{(2\lambda+1)} \sum_{\mu=-\lambda}^{\lambda} Y^\star_{\lambda\mu}(\vartheta_i, \varphi_i)\, Y_{\lambda\mu}(\vartheta_j, \varphi_j)\,. \tag{8.11}$$

Assuming the Gaussian form of the function $V_l(|\mathbf{r}_i - \mathbf{r}_j|)$, Bohr and Mottelsson have made an approximation

$$v_\lambda(r_i, r_j) = -4\kappa_\lambda r_i^\lambda r_j^\lambda\,. \tag{8.12}$$

Therefore, the long-range interaction can be written in separable form, i.e., as the sum of the products of functions, which depends on the coordinates of each of the nucleons only:

$$V_l(|\mathbf{r}_i - \mathbf{r}_j|) = -\sum_{\lambda\mu} \frac{16\pi\kappa_\lambda}{2\lambda+1} \left[r_i^\lambda Y^\star_{\lambda\mu}(\vartheta_i, \varphi_i)\right]\left[r_j^\lambda Y_{\lambda\mu}(\vartheta_j, \varphi_j)\right]. \tag{8.13}$$

The whole long-range interaction (8.8) is the sum over all nucleons

$$\hat{V}_{\text{long}} = \frac{1}{2} \sum_{i,j;i\neq j} V_l(|\mathbf{r}_i - \mathbf{r}_j|). \tag{8.14}$$

The factor of $1/2$ in front of the sum avoids a double counting of the interaction between the same particles.
Denoting

$$q_{\lambda\mu}(i) \equiv 2\sqrt{\frac{4\pi}{2\lambda+1}}\, r_i^\lambda Y_{\lambda\mu}(\vartheta_i, \varphi_i), \tag{8.15}$$

one has

$$V_l(|\mathbf{r}_i - \mathbf{r}_j|) \equiv \sum_{\lambda=0}^{\infty} \kappa_\lambda \sum_{\mu=-\lambda}^{\lambda} q^\star_{\lambda\mu}(i)\, q_{\lambda\mu}(j). \tag{8.16}$$

In this way, the operator responsible for the long-range residual interaction $\hat{V}_{\text{long}}$ is the sum of different multipolarity operators $\hat{V}_\lambda$ corresponding to multipole-multipole forces:

$$\hat{V}_{\text{long}} = \frac{1}{2} \sum_{\lambda=0}^{\infty} \hat{V}_\lambda. \tag{8.17}$$

In the second quantization formalism, the operator $\hat{V}_\lambda$ can be written as the sum of the products of one-body operators

$$\hat{V}_\lambda = -\frac{\kappa_\lambda}{2} \sum_{\mu=-\lambda}^{\lambda} \hat{Q}_{\lambda\mu} \hat{Q}^+_{\lambda\mu}, \tag{8.18}$$

where

$$\hat{Q}_{\lambda\mu} = \sum_{\alpha\beta} \langle\alpha|q_{\lambda\mu}|\beta\rangle a^+_\alpha a_\beta \tag{8.19}$$

is the multipole moment operator with the λ multipolarity.

So, one has:

for $\lambda = 0$ – usually included in average potential,
for $\lambda = 1$ – dipole moment causing a mass center shift of nucleus,
for $\lambda = 2$ – quadrupole moment, causing spheroidal shape of nuclei,
for $\lambda = 3$ – octupole moment, giving pear shapes of nuclei,
for $\lambda = 4$ – hexadecapole moment,
for $\lambda = 5$ – duotriacontapole moment,
for $\lambda = 6$ – tetrahexacontapole moment.

Nucleus shapes corresponding to non-zero moments $Q_{\lambda 0}$ are shown in Fig. 6.3.

In practical applications, the residual multipole-multipole interaction is considered by the Hartree-Fock method, which gives a correspondingly deformed single-particle potential.

8.2 SHORT-RANGE EFFECTS

Interaction with a short range much smaller than the radius of the nucleus in extreme cases can be approximated with the Dirac δ function

$$\hat{V}_{\text{short}} = -\hat{V}_0\,\delta(\mathbf{r}_1 - \mathbf{r}_2), \tag{8.20}$$

where $\delta(\mathbf{r}) = \delta(x)\cdot\delta(y)\cdot\delta(z)$ and $\delta(x) = 0$ for $x \neq 0$ and $\int_{-\infty}^{\infty}\delta(x)dx = 1$.

Bohr and Mottelsson adopted $\hat{V}_0$ in a spin-dependent form

$$V_0 = V_0^{\text{s}}\hat{\Pi}_{12}^{\text{s}} + V_0^{\text{t}}\hat{\Pi}_{12}^{\text{t}}, \tag{8.21}$$

where

$$\hat{\Pi}_{12}^{\text{s}} = \frac{1}{4}(1 - \sigma_1\cdot\sigma_2), \tag{8.22}$$

$$\hat{\Pi}_{12}^{\text{t}} = \frac{3}{4}(3 + \sigma_1\cdot\sigma_2) \tag{8.23}$$

are projection operators for singlet (s) and triplet (t) states.

If one evaluates the matrix element of the interaction $\hat{V}_{\text{short}}$ between two-nucleon states coupled to the total angular momentum J

$$|j^2JM\rangle = \sum_{m+m'=M}(|jm\,jm'|JM)a_{jm}^{+}a_{jm'}^{+}|0\rangle, \tag{8.24}$$

one can show that the matrix element $\langle j^2JM|\hat{V}_{\text{short}}|j^2JM\rangle$ strongly depends on J. It has a maximum value in the state $J = 0$ and decreases when $J = 2,4,\ldots$ becomes larger. The spectrum of ^{40}Ca shown in Fig. 8.2 reminds this situation. Therefore, in a relatively good approximation, one replaces the whole short-range interaction by a *pairing force* , which joins nucleons in pairs. By definition, these forces occur only in the two-nucleon states coupled to the total angular momentum $J = 0$. In other states, one assumes that the matrix element of the pairing force disappears.

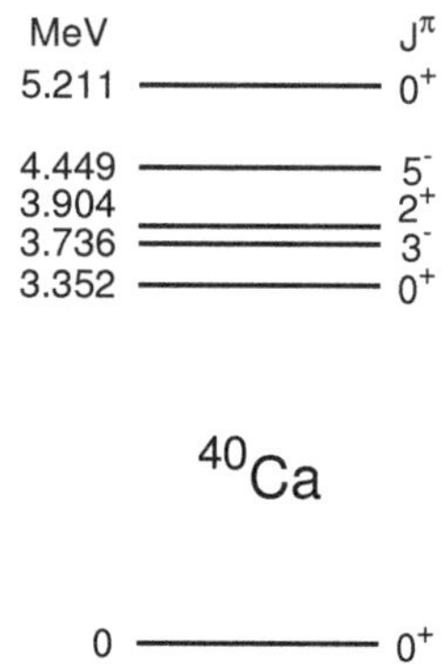

Figure 8.2 Single-particle excited states of the even-even nucleus ^{40}Ca.

In the formalism of the second quantization, the Hamiltonian of such pairing interaction can be written in the following form

$$\hat{H}_{\text{pair}} = -\sum_{j,j'} G_{jj'} \left[a_j^+ a_j^+\right]_0^0 \left[a_{j'} a_{j'}\right]_0^0, \tag{8.25}$$

where

$$\left[a_j^+ a_j^+\right]_0^0 = \sum_{m=-j}^{j} (jm\, j-m|00)\, a_{jm}^+ a_{j-m}^+ \tag{8.26}$$

means the operator of the creation of a pair of particles are coupled to total angular momentum $J = 0$.

Inserting the explicit value of the Clebsch-Gordan coefficient

$$(jm\, j-m|00) = \frac{(-1)^{j-m}}{\sqrt{2j+1}}, \tag{8.27}$$

into Eq. (8.26) one obtains the pairing Hamiltonian in the form

$$\hat{H}_{\text{pair}} = -G \sum_{\substack{j,m>0 \\ j',m'>0}} (-1)^{j-m+j'-m'} a_{jm}^+ a_{j-m}^+ a_{j'-m'} a_{j'm'}, \tag{8.28}$$

where the average pairing strength

$$G \equiv \frac{G_{jj'}}{\sqrt{(2j+1)(2j'+1)}}$$

is usually assumed to be independent of j and j'. Its value depends on the mass region of nuclei and corresponds to the average value of diagonal matrix elements of the short-range interaction

$$G = \overline{\langle \nu, -\nu, |\hat{V}_{\text{short}}| \nu', -\nu' \rangle}, \tag{8.29}$$

where the states

$$|\nu\rangle \equiv |Nljm\rangle \text{ and } |-\nu\rangle \equiv |Nlj-m\rangle,$$

are coupled by the time reversal operation.

8.2.1 Quasispin method

Let us assume the Hamiltonian of the system consisting of the single-particle $\hat{H}_0$ and the pairing parts $\hat{H}_{\text{pair}}$

$$\begin{aligned} \hat{H} = \hat{H}_0 + \hat{H}_{\text{pair}} &= \sum_{j,m>0} e_{jm}(a^+_{jm}a_{jm} + a^+_{j-m}a_{j-m}) \\ &- G \sum_{\substack{j,m>0 \\ j,m'>0}} (-1)^{j-m+j'-m'} a^+_{jm}a^+_{jm}a_{j'-m'}a_{j'm'} . \end{aligned} \tag{8.30}$$

In addition, one assumes that there is one j shell, well separated from the other ones, and the single-particle energies are not dependent on m

$$e_{jm} \equiv e_j \text{ and } \hat{H} \equiv \sum_j \hat{H}_j . \tag{8.31}$$

Let us also introduce the quasispin operators

$$\begin{aligned} \hat{P}^+ &\equiv \sum_{m\rangle 0} (-1)^{j-m} a^+_{jm}a^+_{j-m} , \\ \hat{P} &\equiv \sum_{m\rangle 0} (-1)^{j-m} a_{j-m}a_{jm} , \\ \hat{P}_0 &\equiv \frac{1}{2} \sum_{m\rangle 0} (a^+_{jm}a_{jm} + a^+_{j-m}a_{j-m} - 1) . \end{aligned} \tag{8.32}$$

They obey the commutation rules analogous to those for the angular moment operators $\hat{L}_0, \hat{L}_\pm$

$$[\hat{P}^+, \hat{P}] = 2\hat{P}_0, \quad [\hat{P}_0, \hat{P}^+] = \hat{P}^+, \quad [\hat{P}_0, \hat{P}] = -\hat{P} . \tag{8.33}$$

Therefore, they are called quasispin operators and have similar properties to the angular momentum operators.

Using the definition (8.32), one can write the Hamiltonian (8.30) in the following form:

$$\hat{H}_j = e_j[2\hat{P}_0 + (2j+1)] - G\hat{P}^+\hat{P}\,. \tag{8.34}$$

Let us introduce the quasispin square operator

$$\hat{\mathbf{P}}^2 = \hat{P}^+\hat{P} + \hat{P}_0^2 - \hat{P}_0\,, \tag{8.35}$$

which commutes with all quasispin operators

$$[\hat{\mathbf{P}}^2, \hat{P}^+] = [\hat{\mathbf{P}}^2, \hat{P}] = [\hat{\mathbf{P}}^2, \hat{P}_0] = 0\,.$$

The definition (8.35) enables us to rewrite the Hamiltonian (8.34) in the form

$$\hat{H}_j = e_j[2\hat{P}_0 + \frac{1}{2}(2j+1)] - G(\hat{\mathbf{P}}^2 - \hat{P}_0^2 + \hat{P}_0)\,. \tag{8.36}$$

The whole Hamiltonian is build-up of the $\hat{\mathbf{P}}^2$ and $\hat{P}_0$ operators. So, it commutes with these operators, and it has with them a common set of the $|p, p_0\rangle$ eigenfunctions, where the quantum numbers p and p_0 are the eigenvalues

$$\hat{\mathbf{P}}^2|p\,p_0\rangle = p(p+1)|p\,p_0\rangle\,, \tag{8.37}$$

$$\hat{P}_0\,|p\,p_0\rangle = p_0\,|p\,p_0\rangle\,. \tag{8.38}$$

Similar to angular momentum operators, one has $p_0 = -p, -p+1, \ldots, p$, so $p = -\min(p_0)$ or $p = \max(p_0)$. The eigenproblem of $\hat{H}_j$ (8.34) can be now written in the form:

$$\hat{H}_j\,|p\,p_0\rangle = E_{pp_0}\,|p\,p_0\rangle\,, \tag{8.39}$$

where

$$E_{pp_0} = e_j\left[2p_0 + \frac{1}{2}(2j+1)\right] - G\,[p(p+1) - p_0^2 + p_0]\,. \tag{8.40}$$

What is the physical meaning of quantum numbers: p, p_0 ?
Note that $\hat{P}_0$ is associated with the particle-number operator

$$\hat{N} = \sum_{m\rangle 0}(a^+_{jm}a_{jm} + a^+_{j-m}a_{j-m}) \tag{8.41}$$

by the following relation

$$\hat{P}_0 = \frac{1}{2}\hat{N} - \frac{1}{4}(2j+1)\,. \tag{8.42}$$

Suppose the shell j has n particles, of which $(n-v)$ are coupled to pairs with angular momentum $J=0$. A number v means the number of particles not paired and is called *seniority* . One denotes this state by $|j^n, v\rangle$. The occupation of the j shell is shown graphically in Fig. 8.3, where black spots mark the states occupied by the particles. We have there the $(2j+1)/2$ doubly de-

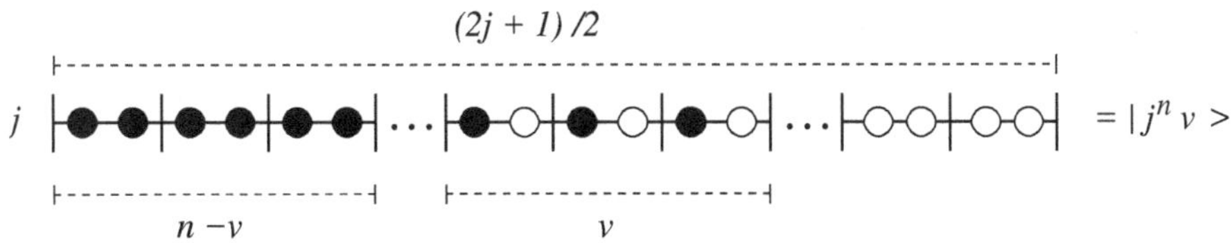

Figure 8.3 Filling $(j, \pm j), (j, \pm|j-1|), \ldots, (j, \pm\frac{1}{2})$ orbitals in the j shell by n particles, including v not paired particles.

generated states occupied by n particles, so

$$\hat{P}_0|j^n v\rangle = \frac{1}{2}\left[n - \frac{1}{2}(2j+1)\right]|j^n v\rangle. \tag{8.43}$$

Hence, using the relation (8.38) one obtains

$$p_0 = \frac{1}{2}\left[n - \frac{1}{2}(2j+1)\right]. \tag{8.44}$$

The operator $\hat{P}$ acting on the state $|j^n v\rangle$ annihilates a pair of particles coupled to $J=0$. After $(n-v)/2$ times repetitions of this operation, all such pairs will disappear, and further action of $\hat{P}$ gives 0. The minimum value of p_0 occurs in the state with $n=v$, i.e., in which there are no more particle pairs coupled to $J=0$

$$\hat{P}_0|j^v v\rangle = \frac{1}{2}\left[v - \frac{1}{2}(2j+1)\right]|j^v v\rangle, \tag{8.45}$$

hence, the total quasispin is

$$p = \frac{1}{2}\left[\frac{1}{2}(2j+1) - v\right]. \tag{8.46}$$

The energy of $\hat{H}_j$ can now be expressed by the angular momentum j of the shell, the total number of particles n on the shell, and the seniority v:

$$\begin{aligned} E_{pp0} &= ne_j - \frac{G}{4}\left\{\left[\frac{1}{2}(2j+1) - v\right]\left[\frac{1}{2}(2j+1) - v + 2\right] + \right. \\ &- \left.\left[n - \frac{1}{2}(2j+1)\right]^2 + \left[n - \frac{1}{2}(2j+1)\right]\right\}. \end{aligned} \tag{8.47}$$

The above result is strict but obtained only for a single j shell. If the particles are spread over several j levels, which is more realistic, other methods must be applied. The quasispin method can be applied to light nuclei where protons and neutrons are on an insulated shell, e.g., at $1f^{7/2}$ orbital. Such a situation happens in nuclei with the proton Z and neutron N numbers between the magic numbers 20 and 28.

CHAPTER 9

Superconducting model of nuclei

A good approximation of the short-range residual forces is the pairing interaction between nucleons of equal but oppositely directed angular momentum. The existence of these forces is evidenced by several experimental data:

- The total angular momentum (spin) of the even-even nucleus is 0, and the odd nucleus spin comes from the angular momentum of the last valence particle.
- Moments of inertia of the nuclei calculated using the average potential only are about three times too large, which indicates correlations between external nucleons.
- Non-magic nuclei should be deformed due to the long-range interactions, while the pairing correlations of outer nucleons stabilize the spherical symmetry of near-magic nuclei.
- Even-even nuclei have an energetic gap between the ground and the first excited non-collective states.

The phenomenon of existence in the nucleus of pairs of nucleons moving in opposite directions is called *superconductivity* by analogy with electrons, which in metals at low temperatures combine in pairs, giving superconducting properties. In 1957, Bardeen, Cooper, and Schrieffer proposed a variational method in the solid-state theory to explain the superconductivity phenomenon, known afterward as the BCS method. One year later, Bohr, Mottelsson, and Pines successfully used this theory to describe the pairing correlations in nuclei.

We will apply the BCS theory to solve appropriately the eigenproblem of a system of fermions in a mean-field potential well interacting by the pairing forces. The following Hamiltonian describes such a system:

$$\hat{H} = \sum_{\nu} e_\nu a_\nu^+ a_\nu - G \sum_{\nu>0,\, \nu'>0} a_\nu^+ a_{-\nu}^+ a_{-\nu'} a_{\nu'} , \tag{9.1}$$

where $|\nu\rangle$ are the single-particle states and the single-particle energies e_ν are the eigenvalues of the single-particle Hamiltonian. $|\nu\rangle$ represents a state with a positive third component of angular momentum here, while $|-\nu\rangle$ is a negative third component of angular momentum. Let us recall here that the Kramers' theorem says that fermionic states are twice degenerated if the Hamiltonian is time reversal invariant. It means the states $|\nu\rangle$ and $|-\nu\rangle$ have the same energy e_ν. The first term of the Hamiltonian (9.1) describes the energy of non-interacting particles, and the second one represents the contribution from pairing interaction, where G is the average strength of these forces (8.29). The fermion creation a_ν^+ and annihilation a_ν operators meet anti-commutation conditions (7.7), (7.10), and (7.12) to take the Pauli principle into account. Let us determine the vacuum of fermions $|0\rangle$ as a state where it is no longer possible to annihilate the particle (7.5)

$$a_\nu |0\rangle = 0 \quad \text{so, also} \quad \langle 0| a_\nu^+ = 0 , \tag{9.2}$$

since the creation operator $a_\nu^+ = (a_\nu)^+$ is Hermitian coupling of the particle annihilation operator.

9.1 EVEN-EVEN NUCLEI

Let us consider the even number n of fermions of one kind. Bohr, Mottelson, and Pines assumed that the BCS function describing the ground state of such a system has the form

$$|\Phi_0\rangle = \prod_{\nu>0} (U_\nu + V_\nu a_\nu^+ a_{-\nu}^+)|0\rangle , \tag{9.3}$$

where U_ν and V_ν are the variational parameters related to the occupation probability of the level e_ν by the pair particles. Normalization of the function $|\Phi_0\rangle$ requires that

$$U_\nu^2 + V_\nu^2 = 1 \tag{9.4}$$

for each state ν. So, the values of coefficients U_ν, V_ν can vary from 0 to 1, and their values are determined by minimizing the system's total energy. These coefficients are equal in the $|\nu\rangle$ and $|-\nu\rangle$ states:

$$U_\nu = U_{-\nu} ; \quad V_\nu = V_{-\nu} . \tag{9.5}$$

The BCS function has an important disadvantage. Namely, it does not correspond to a given number of particles. E.g., for the two states $|1\rangle$ and $|2\rangle$ one has

$$\begin{aligned} |\Phi_0^{12}\rangle &= (U_1 + V_1 a_1^+ a_{-1}^+)(U_2 + V_2 a_2^+ a_{-2}^+)|0\rangle \\ &= U_1 U_2 + U_1 V_2 a_2^+ a_{-2}^+ \\ &\quad + V_1 U_2 a_1^+ a_{-1}^+ + V_1 V_2 a_1^+ a_{-1}^+ a_2^+ a_{-2}^+ |0\rangle . \end{aligned} \tag{9.6}$$

It means that in such a BCS function, there are states with zero, two, and four particles. It is, therefore, necessary to conserve, at least on average, the number of particles n. It is achieved by determining the expected value of the particle number operator (7.15)

$$\hat{N} = \sum_{\nu} a_\nu^+ a_\nu \tag{9.7}$$

in state $|\Phi_0\rangle$. This leads to a variational problem

$$\langle\Phi_0|\hat{H}|\Phi_0\rangle = \text{minimum}, \tag{9.8}$$

where minimization is performed with respect to the parameters U_ν (or V_ν) with an additional condition for the number of particles

$$\langle\Phi_0|\hat{N}|\Phi_0\rangle = n. \tag{9.9}$$

To take this condition into account, one uses the Lagrange multiplier λ technique and minimizes the function

$$\langle\Phi_0|\hat{H}|\Phi_0\rangle - \lambda\langle\Phi_0|\hat{N}|\Phi_0\rangle = \text{minimum}. \tag{9.10}$$

It corresponds to the minimization of the mean value of the following modified Hamiltonian (called *Routhian*):

$$\hat{H}' = \hat{H} - \lambda\hat{N} = \sum_{\nu > 0} (e_\nu - \lambda)(a_\nu^+ a_\nu + a_{-\nu}^+ a_{-\nu}) - G \sum_{\substack{\nu' > 0, \nu \neq \nu' \\ \nu > 0}} a_\nu^+ a_{-\nu}^+ a_{-\nu'} a_{\nu'}, \tag{9.11}$$

in which single-particle energies are calculated relative to the multiplier λ, which is also called the chemical potential or the Fermi level .

It is convenient to introduce a new object called *quasiparticle* in order to evaluate more easily the average values in Eq. (9.10). Namely, one assumes that the BCS ground state $|\Phi_0\rangle$ is the quasiparticle vacuum. This means that the quasiparticle annihilation operator α_k acting of the BCS state gives

$$\alpha_k|\Phi_0\rangle = 0 \text{ and } \langle\Phi_0|\alpha_k^+ = 0. \tag{9.12}$$

The annihilation (α_k) and creation (α_k^+) operators of a quasiparticle, which are fermions, have also to fulfill the anti-commutation rules

$$\left\{\alpha_\nu, \alpha_{\nu'}^+\right\} = \delta_{\nu\nu'}, \quad \{\alpha_\nu, \alpha_{\nu'}\} = \left\{\alpha_\nu^+, \alpha_{\nu'}^+\right\} = 0. \tag{9.13}$$

One of the possible forms of creation and annihilation operators of quasiparticles providing conditions (9.12) and (9.13) gives the Bogolubov-Valatin transformation:

$$\begin{aligned} \alpha_\nu &= U_\nu a_\nu - V_\nu a_{-\nu}^+, \\ \alpha_{-\nu} &= U_\nu a_{-\nu} + V_\nu a_\nu^+. \end{aligned} \tag{9.14}$$

The creation operator of quasiparticles can be obtained by the Hermitian coupling of the above equations

$$\begin{aligned} \alpha_\nu^+ &= U_\nu a_\nu^+ - V_\nu a_{-\nu}, \\ \alpha_{-\nu}^+ &= U_\nu a_{-\nu}^+ + V_\nu a_\nu. \end{aligned} \tag{9.15}$$

The inverse Bogolubov-Valatin transformation is needed for further calculations

$$\begin{aligned} a_\nu &= U_\nu \alpha_\nu + V_\nu \alpha_{-\nu}^+, \\ a_{-\nu} &= U_\nu \alpha_{-\nu} - V_\nu \alpha_\nu^+, \\ a_\nu^+ &= U_\nu \alpha_\nu^+ + V_\nu \alpha_{-\nu}, \\ a_{-\nu}^+ &= U_\nu \alpha_{-\nu}^+ - V_\nu \alpha_\nu. \end{aligned} \tag{9.16}$$

Using the above transformation, one can easily calculate the mean value of the Hamiltonian (9.11) in the BCS ground-state $|\Phi_0\rangle$

$$\begin{aligned} \langle\Phi_0|\hat{H}'|\Phi_0\rangle &= \langle\Phi_0|\sum_{\nu>0}(e_\nu - \lambda)\{(U_\nu\alpha_\nu^+ + V_\nu\alpha_{-\nu})(U_\nu\alpha_\nu + V_\nu\alpha_{-\nu}^+) \\ &+ (U_\nu\alpha_{-\nu}^+ - V_\nu\alpha_\nu)(U_\nu\alpha_{-\nu} - V_\nu\alpha_\nu^+)\} \\ &- G\sum_{\nu,\nu'>0}(U_\nu\alpha_\nu^+ + V_\nu\alpha_{-\nu})(U_\nu\alpha_{-\nu}^+ - V_\nu\alpha_\nu) \\ &\cdot\left(U_{\nu'}\alpha_{-\nu'} - V_{\nu'}\alpha_{\nu'}^+\right)\left(U_{\nu'}\alpha_{\nu'} + V_{\nu'}\alpha_{-\nu'}^+\right)|\Phi_0\rangle. \end{aligned} \tag{9.17}$$

Then by shifting the operators α_ν to the right hand side and α_ν^+ to the left and using the anti-commutation rules (9.13) and Eq. (9.12), one obtains the following formula:

$$\langle\Phi_0|\hat{H}'|\Phi_0\rangle = 2\sum_{\nu>0}(e_\nu - \lambda)V_\nu^2 - G\sum_{\nu>0}V_\nu^4 - G\left(\sum_{\nu>0}U_\nu V_\nu\right)^2. \tag{9.18}$$

This expression reaches the minimal value when its variation over V_ν (or U_ν) disappears, i.e.

$$\sum_\nu \frac{\partial}{\partial V_\nu}\langle\Phi_0|\hat{H}'|\Phi_0\rangle\delta V_\nu = 0. \tag{9.19}$$

There must be a condition for every V_ν

$$\frac{\partial}{\partial V_\nu}\langle\Phi_0|\hat{H}'|\Phi_0\rangle = 0. \tag{9.20}$$

The derivative of the expression (9.18) over V_ν is:

$$4(e_\nu - \lambda)V_\nu - 4G\,V_\nu^3 - 2G\left(\sum_{\mu>0}U_\mu V_\mu\right)\left(U_\nu + V_\nu\frac{\partial U_\nu}{\partial V_\nu}\right) = 0. \tag{9.21}$$

The BCS function normalization condition (9.4) leads to the relation

$$\frac{\partial U_\nu}{\partial V_\nu} = -\frac{V_\nu}{U_\nu}. \tag{9.22}$$

Dividing both sides of (9.21) by 2 and using Eq. (9.22), one obtains the following relationship:

$$2(e_\nu - \lambda)V_\nu - 2GV_\nu^3 - G\left(\sum_{\mu>0}U_\mu V_\mu\right)\left(U_\nu - \frac{V_\nu^2}{U_\nu}\right) = 0. \tag{9.23}$$

Let us introduce a parameters Δ called the *energy-gap*

$$\Delta \equiv G\sum_{\mu>0}U_\mu V_\mu \tag{9.24}$$

and shifted single-particle energies (with a small correction coming from pairing forces) measured with respect to the Fermi level

$$\tilde{e}_\nu \equiv e_\nu - \lambda - GV_\nu^2. \tag{9.25}$$

Then multiplying Eq. (9.23) by U_ν, one gets

$$2\tilde{e}_\nu V_\nu U_\nu - \Delta(U_\nu^2 - V_\nu^2) = 0. \tag{9.26}$$

From Eqs. (9.26) and (9.4) one can obtain expressions for U_ν^2 and V_ν^2:

$$U_\nu^2 = \frac{1}{2}\left(1 + \frac{\tilde{e}_\nu}{\sqrt{\tilde{e}_\nu^2 + \Delta^2}}\right), \quad V_\nu^2 = \frac{1}{2}\left(1 - \frac{\tilde{e}_\nu}{\sqrt{\tilde{e}_\nu^2 + \Delta^2}}\right), \tag{9.27}$$

which describe the probabilities of occupying single-particle states and depend on the parameters λ (see Eq. 9.10) and Δ. It is easy to evaluate the average number of particles in BCS state when one expresses the particle number operator (9.7) by the quasiparticle operators: (9.16)

$$\begin{aligned} n &= \langle\Phi_0|\hat{N}|\Phi_0\rangle = \langle\Phi_0|\sum_{\nu>0}\{(U_\nu\alpha_\nu^+ + V_\nu\alpha_{-\nu})(U_\nu\alpha_\nu + V_\nu\alpha_{-\nu}^+) \\ &+ (U_\nu\alpha_{-\nu}^+ - V_\nu\alpha_\nu)(U_\nu\alpha_{-\nu} - V_\nu\alpha_\nu^+)\}|\Phi_0\rangle \\ &= 2\sum_{\nu>0} V_\nu^2 . \end{aligned} \tag{9.28}$$

Hence, after inserting Eq. (9.27) into (9.28) one has

$$\sum_{\nu>0}\left(1 - \frac{\tilde{e}_\nu}{\sqrt{\tilde{e}_\nu^2 + \Delta^2}}\right) = n. \tag{9.29}$$

This equation is used to determine λ. The value of the energy gap one obtains by inserting the explicit form U_ν and V_ν into the definition of Δ (9.24)

$$\Delta = \frac{G}{2}\sum_{\nu>0}\frac{\Delta}{\sqrt{\tilde{e}_\nu^2 + \Delta^2}}. \tag{9.30}$$

The last equation, after dividing both sides by $\Delta G/2$, takes the form:

$$\frac{2}{G} = \sum_{\nu>0}\frac{1}{\sqrt{\tilde{e}_\nu^2 + \Delta^2}}. \tag{9.31}$$

After omission of small terms $GV_\nu^2 \ll e_\nu - \lambda$ in the above equations, one obtains the following set of the BCS equations:

$$U_\nu^2 = \frac{1}{2}\left(1 + \frac{e_\nu - \lambda}{\sqrt{(e_\nu - \lambda)^2 + \Delta^2}}\right),$$

$$U_\nu^2 + V_\nu^2 = 1\,,$$
$$\sum_{\nu>0}\left(1 - \frac{e_\nu - \lambda}{\sqrt{(e_\nu-\lambda)^2+\Delta^2}}\right) = n\,, \tag{9.32}$$
$$\frac{2}{G} = \sum_{\nu>0}\frac{1}{\sqrt{(e_\nu-\lambda)^2+\Delta^2}}\,.$$

The above system of coupled equations one solves numerically by iteration.

Having the values Δ and λ, one can evaluate the ground state energy using the formula:

$$E_0 = 2\sum_{\nu>0} e_\nu V_\nu^2 - G\sum_{\nu>0} V_\nu^4 - \frac{\Delta^2}{G} \tag{9.33}$$

and obtain the ground-state function $|\Phi_0\rangle$ by Eq. (9.3).

9.2 DISCUSSION OF THE BCS EQUATIONS

To understand the physical meaning of the Δ, λ, U_ν and V_ν parameters, let us discuss in detail the set of equations (9.32).

The quantity directly related to the pairing interaction is Δ. Note that when $\Delta = 0$, the BCS equations describe the system of free fermions. Parameters U_ν and V_ν then have fixed values

$$U_\nu^2 = \frac{1}{2}\left(1 + \frac{(e_\nu-\lambda)}{\sqrt{(e_\nu-\lambda)^2}}\right) = \frac{1}{2}(1\pm 1) = \begin{cases} 1 & \text{for } e_\nu > \lambda, \\ 0 & \text{for } e_\nu < \lambda, \end{cases}$$
$$\tag{9.34}$$
$$V_\nu^2 = \frac{1}{2}\left(1 - \frac{e_\nu-\lambda}{\sqrt{(e_\nu-\lambda)^2}}\right) = \frac{1}{2}(1\mp 1) = \begin{cases} 0 & \text{for } e_\nu > \lambda, \\ 1 & \text{for } e_\nu < \lambda. \end{cases}$$

So, the probability of filling the states having energy smaller than λ by a pair of particles is equal to 1, while for those lying above λ is equal to 0. It means that λ plays the role of the Lagrange multiplier and simultaneously is the system's Fermi energy.

For $\Delta = 0$, the Bogolubov-Valatin transformation is reduced to

$$\alpha_\nu = \begin{cases} a_\nu & \text{for } e_\nu > \lambda, \\ -a_{-\nu}^+ & \text{for } e_\nu < \lambda, \end{cases}$$
$$\tag{9.35}$$
$$\alpha_\nu^+ = \begin{cases} a_\nu^+ & \text{for } e_\nu > \lambda, \\ -a_{-\nu} & \text{for } e_\nu < \lambda. \end{cases}$$

Then, the quasiparticle annihilation operators are the annihilation operators of particles above the Fermi level and the creation operator below λ. The quasiparticle creation operators create particles in empty states and annihilate them in the occupied ones.

The equation for the number of particles for $\Delta = 0$ takes the form

$$n = \sum_{\nu>0}\left(1 - \frac{e_\nu - \lambda}{\sqrt{(e_\nu - \lambda)^2}}\right) = \sum_{\nu>0}^{\Omega}(1+1) = 2\Omega \tag{9.36}$$

and the number of particles equals twice the number of Ω states below the Fermi level. The ground state energy in such a case is

$$E_0 = 2\sum_{\nu>0}^{\Omega} e_\nu V_\nu^2 = 2\sum_{\nu>0}^{\Omega} e_\nu \tag{9.37}$$

and equals twice the sum of single-particle energies below the Fermi level.

The above observations prove that in the case of $\Delta = 0$, superconductivity is not observed, the pairing of nucleons does not occur, and the nucleus behaves like a system of free fermions in the average potential. The solution of superconductivity equations, which leads to $\Delta = 0$, is called normal - no pairing effect then exists.

Let us find a condition for the pairing strength that leads to a superconducting solution $\Delta \neq 0$. Inserting to the gap equation (9.31) the following inequality:

$$\sqrt{(e_\nu - \lambda)^2 + \Delta^2} > |e_\nu - \lambda| , \tag{9.38}$$

one can find the upper limit for the value of $2/G$

$$\frac{2}{G} = \sum_{\nu>0} \frac{1}{\sqrt{(e_\nu - \lambda)^2 + \Delta^2}} < \sum_{\nu>0} \frac{1}{|e_\nu - \lambda|} . \tag{9.39}$$

Hence, one has

$$G > 2\left(\sum_{\nu>0} \frac{1}{|e_\nu - \lambda|}\right)^{-1} \equiv G_{\mathrm{cr}} . \tag{9.40}$$

In BCS theory, for the pairing strength less than the critical value G_{cr}, no superconducting solution occurs and $\Delta = 0$.

The energy dependence of the occupation probability V_ν^2 in the superconducting case is shown in Fig. 9.1. If single-particle energy is much smaller than the Fermi energy $e_\nu \ll \lambda$, then

$$V_\nu^2 \simeq 1, \quad U_\nu^2 \simeq 0.$$

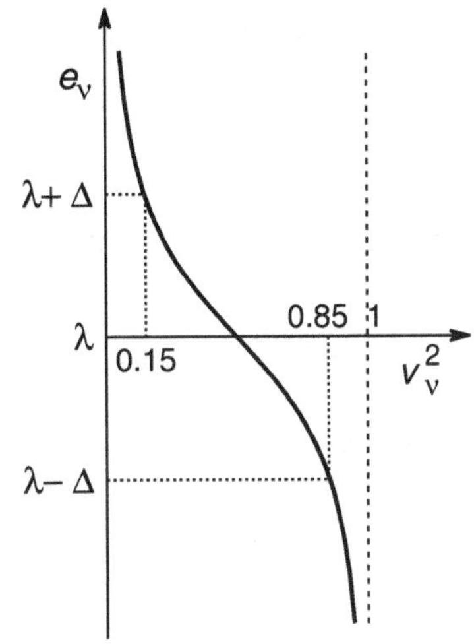

Figure 9.1 Occupation probability of single-particle levels around the Fermi surface in the superconducting BCS model.

In the state with $e_\nu = \lambda$ one has $V_\nu^2 = \frac{1}{2}$ and $U_\nu^2 = \frac{1}{2}$ while for energies much larger than the Fermi energy ($e_\nu \gg \lambda$) one has

$$V_\nu^2 \simeq 0, \quad U_\nu^2 \simeq 1 .$$

As can be seen, far from the Fermi surface, the coefficients U_ν and V_ν have the same values as for a free fermions system. One can see that smoothing the occupation probability of the single-particle states occurs within a relatively narrow energy window $(\lambda - \Delta, \lambda + \Delta)$:

$$V_\nu^2 = \frac{1}{2}\left(1 - \frac{\Delta}{\Delta\sqrt{2}}\right) \simeq 0.15 \ \text{ for } \ e_\nu = \lambda + \Delta , \tag{9.41}$$

$$V_\nu^2 = \frac{1}{2}\left(1 + \frac{\Delta}{\Delta\sqrt{2}}\right) \simeq 0.85 \ \text{ for } \ e_\nu = \lambda - \Delta . \tag{9.42}$$

This means that the sharp Fermi level for free fermions is smoothed in the superconducting case within the interval approximately equal to 2Δ.

9.3 FREE QUASIPARTICLES METHOD

Bogolubov's variational procedure leading to the system of BCS equations is equivalent to the assumption that quasiparticles do not interact with each other. It means the terms responsible for two-quasiparticle interactions, which contain products of four operators α_ν, α_ν^+ can be neglected.

Let us express again the Hamiltonian (9.1) by the quasiparticle operators and bring it to the normal form, wherein annihilation operators are on the right and creation on the left side. Hamiltonian can be represented by the following sum of operators $\hat{H}_{ij}$:

$$\hat{H} = \hat{H}_{00} + \hat{H}_{11} + \hat{H}_{02} + \hat{H}_{20} + \hat{H}_{22} + \hat{H}_{31} + \hat{H}_{13} + \hat{H}_{40} + \hat{H}_{04}\,, \tag{9.43}$$

where indexes i and j denote the number of the creation and the annihilation operators, respectively, and

$$\begin{aligned}
\hat{H}_{00} &= 2\sum_{\nu>0}\left(e_\nu - GV_\nu^2\right)V_\nu^2 - G\left(\sum_{\nu'>0}U_{\nu'}V_{\nu'}\right)^2 + G\sum_{\nu>0}V_\nu^4 \equiv E_0\,,\\
\hat{H}_{11} &= \sum_{\nu>0}\left\{\left(e_\nu - GV_\nu^2\right)(U_\nu^2 - V_\nu^2) + 2G\left(\sum_{\nu'>0}U_{\nu'}V_{\nu'}\right)U_\nu V_\nu\right\}\\
&\quad\cdot(\alpha_\nu^+\alpha_\nu + \alpha_{-\nu}^+\alpha_{-\nu})\,,\\
\hat{H}_{20} &+ \hat{H}_{02} = 2\sum_{\nu>0}\left\{(e_\nu - GV_\nu^2)U_\nu V_\nu - G\left(\sum_{\nu'>0}U_{\nu'}V_{\nu'}\right)(U_\nu^2 - V_\nu^2)\right\}\\
&\quad\cdot(\alpha_\nu^+\alpha_{-\nu}^+ + \alpha_{-\nu}\alpha_\nu)\,.
\end{aligned}$$

Other terms containing four operators describing the interaction of quasiparticles:

$$\hat{H}_{22} + \hat{H}_{31} + \hat{H}_{13} + \hat{H}_{40} + \hat{H}_{04} \approx 0\,, \tag{9.44}$$

will be omitted. In addition, one demands that $\hat{H}_{20} + \hat{H}_{02} = 0$ and hence one gets the following equation for U_ν and V_ν:

$$2U_\nu V_\nu(e_\nu - GV_\nu^2) - \Delta(U_\nu^2 - V_\nu^2) = 0\,, \tag{9.45}$$

which is the same as the result of minimizing energy (9.26) if single-particle energies are calculated relative to $\lambda : e_\nu \to e_\nu - \lambda \equiv \tilde{e}_\nu$. So, the condition (9.44) leads to the same system of BCS equations as Eq. (9.32).

The Hamiltonian (9.43) takes now a very simple form

$$\hat{H} = E_0 + \sum_\nu E_\nu \alpha_\nu^+\alpha_\nu\,, \tag{9.46}$$

where $E_\nu = \sqrt{\tilde{e}_\nu^2 + \Delta^2}$ is the quasiparticle energy and E_0 is given by Eq. (9.33).

9.4 EXCITED STATES OF EVEN NUCLEI

The excitation of nuclei with an even number of nucleons occurs by breaking a pair of particles. At the same time, the excited wave function should contain only components with an even number of particles, so it can only be two-quasiparticle excitations

$$|\Phi_{12}\rangle = \alpha^+_{\nu_1}\alpha^+_{\nu_2}|\Phi_0\rangle. \tag{9.47}$$

The energy of the system in the above state is

$$\begin{aligned} E_{12} &= \langle\Phi_{12}|\hat{H}|\Phi_{12}\rangle = \langle\Phi_0|\alpha_{\nu_2}\alpha_{\nu_1}\left[E_0 + \sum_\nu E_\nu \alpha^+_\nu \alpha_\nu\right]\alpha^+_{\nu_1}\alpha^+_{\nu_2}|\Phi_0\rangle \\ &= E_0 + E_{\nu_1} + E_{\nu_2}. \end{aligned} \tag{9.48}$$

The excitation energy W corresponding to the state $|\Phi_{12}\rangle$ is therefore:

$$W = E_{\nu_1} + E_{\nu_2} = \sqrt{\tilde{e}^2_{\nu_1} + \Delta^2} + \sqrt{\tilde{e}^2_{\nu_2} + \Delta^2}. \tag{9.49}$$

It is always greater than 2Δ and provides an energy gap between the ground and the first excited state of the even-even nucleus. The lowest excited state one obtains by creating a pair of quasiparticles at the single-particle levels closest to the Fermi energy $e_{\nu_1} \approx e_{\nu_2} \approx \lambda$. Then $E_{\nu_1} \approx E_{\nu_2} \approx \Delta$ and the lowest excited state lies 2Δ above the ground state.

The same system of BCS equations was used to describe the excited and ground states. This is an approximation. In the next section, one will check whether it is good or not.

9.5 BLOCKING EFFECT

Let us derive the equations of superconductivity using the variational method for the two-particle state $|\Phi_{12}\rangle = \alpha^+_{\nu_1}\alpha^+_{\nu_2}|\Phi_0\rangle$. For this purpose, one minimizes the functional:

$$\begin{aligned} \langle\Phi_{12}|\hat{H} - \lambda\hat{N}|\Phi_{12}\rangle &= \langle\Phi_0|\alpha_{\nu_2}\alpha_{\nu_1}(\hat{H} - \lambda\hat{N})\alpha^+_{\nu_1}\alpha^+_{\nu_2}|\Phi_0\rangle \\ &= \sum_{\nu>0} 2V^2_\nu \tilde{e}_\nu - G\left(\sum_{\nu'>0} U_{\nu'}V_{\nu'}\right)^2 \\ &\quad + G\sum_{\nu>0} V^4_\nu \\ &\quad + \left(U^2_{\nu_1} - V^2_{\nu_1}\right)\tilde{e}_{\nu_1} + 2G\left(\sum_{\nu'>0} U_{\nu'}V_{\nu'}\right)U_{\nu_1}V_{\nu_1} \\ &\quad + \left(U^2_{\nu_2} - V^2_{\nu_2}\right)\tilde{e}_{\nu_2} + 2G\left(\sum_{\nu'>0} U_{\nu'}V_{\nu'}\right)U_{\nu_2}V_{\nu_2}, \end{aligned} \tag{9.50}$$

with respect to V_ν, where $\nu \neq \nu_1$ and $\nu \neq \nu_2$. Therefore

$$\frac{\partial \langle \Phi_{12} | \hat{H} - \lambda \hat{N} | \Phi_{12} \rangle}{\partial V_\nu} = 0. \tag{9.51}$$

Hence, one gets the equation for each state ν

$$\begin{aligned} &4V_\nu \tilde{e}_\nu - 4GV_\nu^3 - 2G\left(\sum_{\nu'>0} U_{\nu'} V_{\nu'}\right)\left(U_\nu - \frac{V_\nu^2}{U_\nu}\right) \\ &+2G\left(U_\nu - \frac{V_\nu^2}{U_\nu}\right) U_{\nu_1} V_{\nu_1} + 2G\left(U_\nu - \frac{V_\nu^2}{U_\nu}\right) U_{\nu_2} V_{\nu_2} = 0. \end{aligned} \tag{9.52}$$

Multiplying them on both sides by $U_\nu/2$ one gets:

$$2U_\nu V_\nu \tilde{e}_\nu - \left(G \sum_{\substack{\nu'>0 \\ \nu' \neq \nu_1, \nu' \neq \nu_2}} U_{\nu'} V_{\nu'} \right) \left(U_\nu^2 - V_\nu^2 \right) = 0. \tag{9.53}$$

Let us introduce the energy gap $\Delta_{\nu_1 \nu_2}$ in a system with 'blocked' states ν_1 and ν_2

$$\Delta_{\nu_1 \nu_2} \equiv G \sum_{\substack{\nu>0 \\ \nu \neq \nu_1, \nu \neq \nu_2}} U_\nu V_\nu . \tag{9.54}$$

The coefficients U_ν and V_ν are now described by equations

$$U_\nu^2 = \frac{1}{2}\left(1 + \frac{\tilde{e}_\nu}{\sqrt{\tilde{e}_\nu^2 + \Delta_{\nu_1 \nu_2}^2}} \right), \quad V_\nu^2 = \frac{1}{2}\left(1 - \frac{\tilde{e}_\nu}{\sqrt{\tilde{e}_\nu^2 + \Delta_{\nu_1 \nu_2}^2}} \right). \tag{9.55}$$

Particle number equation

$$\langle \Phi_{12} | \hat{N} | \Phi_{12} \rangle = n \tag{9.56}$$

takes the form

$$2 + \sum_{\substack{\nu>0 \\ \nu \neq \nu_1, \nu \neq \nu_2}} \left(1 + \frac{\tilde{e}_\nu}{\sqrt{\tilde{e}_\nu^2 + \Delta_{\nu_1 \nu_2}^2}} \right) = n, \tag{9.57}$$

and the energy gap equation is

$$\frac{2}{G} = \sum_{\substack{\nu>0 \\ \nu \neq \nu_1, \nu \neq \nu_2}} \left(\frac{1}{\sqrt{\tilde{e}_\nu^2 + \Delta_{\nu_1 \nu_2}^2}} \right). \tag{9.58}$$

The energy of the two-particle $|\Phi_{12}\rangle$ is described by the equation:

$$\begin{aligned} E_{12} &= \langle\Phi_{12}|\hat{H}|\Phi_{12}\rangle = e_{\nu_1} + e_{\nu_2} + 2 \sum_{\substack{\nu>0 \\ \nu\neq\nu_1,\nu\neq\nu_2}} e_\nu V_\nu^2 - \frac{\Delta^2_{\nu_1\nu_2}}{G} \\ &- G \sum_{\substack{\nu>0 \\ \nu\neq\nu_1,\nu\neq\nu_2}} V_\nu^4 . \end{aligned} \tag{9.59}$$

As one can see, all states occupied by quasiparticles $|\nu_1\rangle$ and $|\nu_2\rangle$ are excluded from the sums in the above superconductivity equations. These states are blocked for pairing interactions and are not subject to the smoothed occupation probability of these states. This reflects the form of the wave function

$$\begin{aligned} |\Phi_{12}\rangle &= \alpha^+_{\nu_1}\alpha^+_{\nu_2}|\Phi_0\rangle = \\ &= \left(U_{\nu_1}a^+_{\nu_1} + V_{\nu_1}a_{-\nu_1}\right)\left(U_{\nu_2}a^+_{\nu_2} + V_{\nu_2}a_{-\nu_2}\right)|\Phi_0\rangle \\ &= a^+_{\nu_1}a^+_{\nu_2} \prod_{\substack{\nu\neq\nu_1,\nu>0 \\ \nu\neq\nu_2}} (U_\nu + V_\nu a^+_\nu a^+_{-\nu})|0\rangle , \end{aligned} \tag{9.60}$$

in which the states ν_1 and ν_2 are filled by single particles with a probability of 1 and the other states by pairs of particles with probability V_ν^2. The blocking effect in the two-quasiparticle state is illustrated in Fig. 9.2. The ν_1 and ν_2 states are occupied by single nucleons.

For many particles in the nucleus, the blocking effect does not play a significant role and is often not considered. The blocking complicates the excited state wave function's construction, as such functions corresponding to different excited states are not orthogonal. One has to perform a rather complicated mathematical procedure to construct an orthogonal set of functions from them. The blocking effect can be significant when the pairing strength is close to critical. One obtains better and more accurate results when one projects the BCS wave function into a good number of particles, but such a procedure is even more challenging to apply.

9.6 ODD NUCLEI

The ground and excited state of systems with an odd number of particles may be described by a BCS wave functions in which all single-particle states, except one in which a single nucleon or a hole occurs, are occupied with the BCS probability by pairs of particles. In general, the function of such a

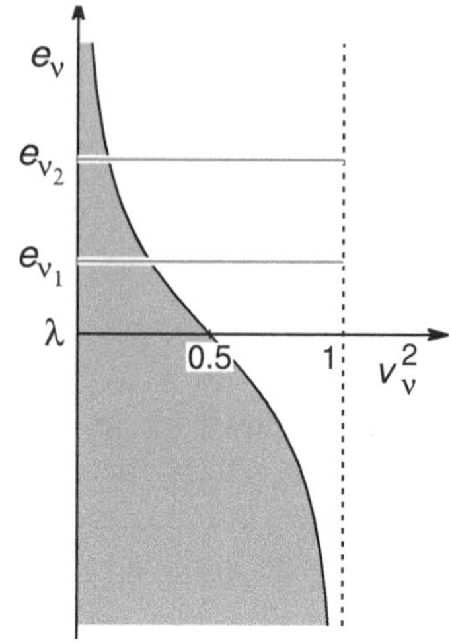

Figure 9.2 Excited states of the even nuclei in the BCS model.

system with an odd number of nucleons is described by the function:

$$|\Phi_1^{\text{odd}}\rangle = \alpha_{\nu_1}^{+}|\Phi_0\rangle. \tag{9.61}$$

The creation of a quasiparticle in the state of ν_1 contributes to ground state energy of an even system by an additional energy E_{ν_1}

$$E_1^{\text{odd}} = \langle\Phi_0|\alpha_{\nu_1}\hat{H}\alpha_{\nu_1}^{+}|\Phi_0\rangle = E_0 + E_{\nu_1}, \tag{9.62}$$

which is equal to the quasiparticle energy when one does not consider the blocking effect. The lowest energy of the odd-particle system occurs in a situation when the quasiparticle is created in a state closest to Fermi's level, e.g., in $e_{\nu_0} \approx \lambda$ and $\alpha_{\nu_0}^{+}|\Phi_0\rangle$. Then

$$E_0^{\text{odd}} = \langle\Phi_0|\alpha_{\nu_0}\hat{H}\alpha_{\nu_0}^{+}|\Phi_0\rangle \approx E_0 + \Delta. \tag{9.63}$$

The excitation energy W of the odd-nucleus can be any small as:

$$W = \sqrt{\tilde{e}_{\nu_1}^2 + \Delta^2} - \Delta < |\tilde{e}_{\nu_1}|. \tag{9.64}$$

It means that the pairing forces effectively make nearer single-particle energy levels.

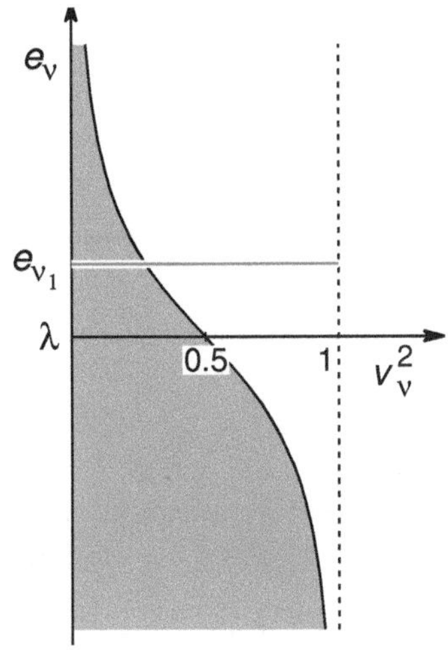

Figure 9.3 Excited states of the odd nuclei in the BCS model.

As a result of minimizing the energy of the odd-system relative to V_ν ($\nu \neq \nu_1$) one obtains a system of the BCS equations in which the state ν_1 has been blocked:

$$U_\nu^2 = \frac{1}{2}\left(1 + \frac{\tilde{e}_\nu}{\sqrt{\tilde{e}_\nu^2 + \Delta_{\nu_1}^2}}\right), \quad V_\nu^2 = \frac{1}{2}\left(1 - \frac{\tilde{e}_\nu}{\sqrt{\tilde{e}_\nu^2 + \Delta_{\nu_1}^2}}\right) \tag{9.65}$$

$$n = 1 + 2\sum_{\substack{\nu>0 \\ \nu\neq\nu_1}}\left(1 - \frac{\tilde{e}_\nu}{\sqrt{\tilde{e}_\nu^2 + \Delta_{\nu_1}^2}}\right), \tag{9.66}$$

$$\frac{2}{G} = \sum_{\substack{\nu>0 \\ \nu\neq\nu_1}}\left(\frac{1}{\sqrt{\tilde{e}_\nu^2 + \Delta_{\nu_1}^2}}\right), \quad \text{where} \quad \Delta_{\nu'} \equiv G\sum_{\substack{\nu>0 \\ \nu\neq\nu_1}} U_\nu V_\nu . \tag{9.67}$$

In the state ν_1, there is a free particle or hole, and the smoothed occupation probability runs across all other levels, as it is shown in Fig. 9.3.
The energy of the system with the odd number of particles is

$$E_1^{\text{odd}} = e_{\nu_1} + 2\sum_{\substack{\nu>0 \\ \nu\neq\nu_1}} V_\nu^2 e_\nu - \frac{\Delta_{\nu_1}^2}{G} - G\sum_{\substack{\nu>0 \\ \nu\neq\nu_1}} V_\nu^4 . \tag{9.68}$$

This result is slightly different from expression (9.62) because it includes blocking of ν_1 by an odd nucleon.

CHAPTER 10

Deformed nuclei

This chapter will show how to calculate the potential energy of deformed atomic nuclei in a more precise macro-micro method than the one of Myers-Swiatecki presented in Chapter 5. One has already discussed the macroscopic model in which strong correlations between nucleons, like in a liquid drop, are present. Another microscopic model, the single-particle approach, assumes only free particles moving in the average potential field. The idea is to extract shell effects from average nuclear properties and describe, at least approximately using the BCS method, the effect of short-range pairing interaction. The effect of long-range forces is included, as shown in Sec. 8.1, by introducing an additional collective coordinate: nuclear deformation. The minimum of the total nuclear energy with respect to the deformation parameters determines the shape of the nucleus in the ground state and it is called equilibrium deformation.

Since one does not know initially the equilibrium deformation of the nucleus corresponding to its minimal energy, calculations have to be made on a grid in the deformation parameter space. The deformation parameters are parametric variables against which the energy is minimized.

One can evaluate the macroscopic part of the nuclear energy using, e.g., the liquid-drop formula (5.43), which, after a small modification, can be written as

$$\begin{aligned} E_{\rm LD}(N,Z,{\rm def}) &= a_1(1-\kappa I^2)A + a_2(1-\kappa I^2)A^{2/3}\,B_{\rm s}({\rm def}) \\ &+ \frac{3}{5}\frac{Z^2e^2}{r_0}A^{-1/3}B_{\rm c}({\rm def})\,, \end{aligned} \tag{10.1}$$

where $I = \frac{N-Z}{A}$, and parameters a_1, a_2, κ and r_0 are given in Eq. (5.61).

One can also evaluate macroscopic energy within the droplet model (5.71) which is an extended version of the liquid drop model, or using the

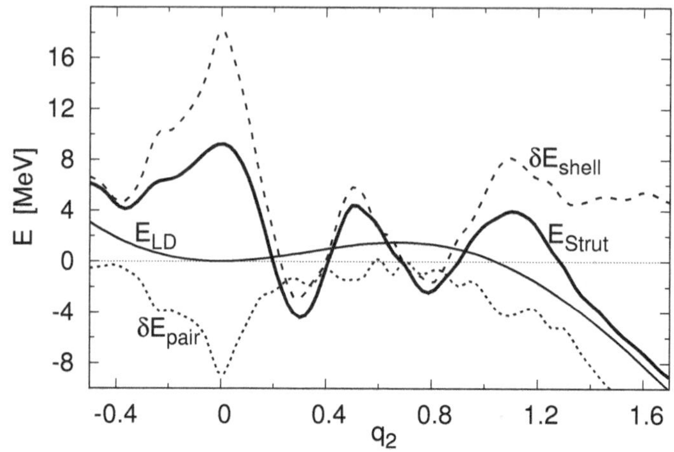

Figure 10.1 Different terms of the potential energy of the nucleus ^{252}Cf as a function of elongation of the nucleus: $E_{\rm LD}$ — macroscopic drop energy (thin solid line), $\delta E_{\rm shell}$ — shell correction (dashed line), $\delta E_{\rm pair}$ — pairing correction (dotted line), $E_{\rm Strut} = E_{\rm LD} + \delta E_{\rm shell}, +\delta E_{\rm pair}$ — total Strutinsky energy (thick solid line).

Yukawa-folded method (5.76). The macroscopic energy of non-rotating nuclei always reaches a minimum at the spherical shape, as it is visible, e.g., in Fig. 10.1, where the liquid drop energy ($E_{\rm LD}$, thin solid line) of ^{252}Cf is plotted as a function of elongation of a nucleus (q_2). The deformation of a nucleus can only originate from microscopic (quantum mechanical) effects.

10.1 SUMMATION OF SINGLE-PARTICLE ENERGIES

As a result of shell effects, nuclei often exhibit non-spherical shapes at ground state. So, the microscopic contribution to their potential energy has to be calculated using the single-particle energies of the deformed mean-field potentials.

In the first attempt, Mottelson and Nilsson proposed the evaluate the sum of energies of occupied single-particle states and the Coulomb energy

$$E_{\rm MN}(N,Z,{\rm def}) = \sum_{\rm occ} {\rm e}^{\rm p}_{\nu} + \sum_{\rm occ} {\rm e}^{\rm n}_{\nu} + E_{\rm Coul}\,. \tag{10.2}$$

Here e_ν are single-particle energies of protons (p) and neutrons (n) obtained by the diagonalization of the Hamiltonian with the deformed mean-field potential $V(\mathbf{r};{\rm def})$

$$\hat{h} = \hat{T} + V(\mathbf{r};{\rm def})\,. \tag{10.3}$$

The Coulomb energy of protons can be obtained by the integral after all single-particle coordinates of protons

$$E_{\mathrm{Coul}} = \sum_{i>j} \int \int \cdots \int \Psi^*(\mathbf{r}_1,\ldots,\mathbf{r}_z) \frac{\mathrm{e}^2}{|\mathbf{r}_i - \mathbf{r}_j|} \Psi(\mathbf{r}_1,\ldots,\mathbf{r}_z) d\tau_1 \ldots d\tau_z, \quad (10.4)$$

where Ψ is a Slater determinant built of single-particle functions. Replacing the squares of these functions with single-particle proton densities ρ_i^{p}, one gets the formula

$$E_{\mathrm{Coul}} = \sum_{i>j} \int \int \rho_i^{\mathrm{p}}(\mathbf{r}_i) \rho_j^{\mathrm{p}}(\mathbf{r}_j) \frac{\mathrm{e}^2}{|\mathbf{r}_i - \mathbf{r}_j|} d\tau_i d\tau_j. \quad (10.5)$$

Becker and Bethe have estimated this integral as

$$E_{\mathrm{Coul}} = \frac{3}{5} \frac{Z^2 \mathrm{e}^2}{R_0} B_{\mathrm{c}}(\mathrm{def}) - \frac{\pi^2}{2} \frac{5}{3} \frac{a}{R_0^2} - \frac{0.7636}{Z^{2/3}}, \quad (10.6)$$

where a is the value of the diffuseness of the nucleus surface and $R_0 = r_0 A^{1/3}$ is the radius of the spherical charge distribution.

The method of summing single-particle energy gives the minimum energy for non-zero deformations, but it does reproduce neither experimental masses of nuclei nor their fission barrier heights. Its main fault was that the nucleon-nucleon interactions were counted in total energy twice. It is impossible to separate a part of energy originating from the mean-field potential in the energy sum (10.2). Therefore, one has to find another method of including the shell effects.

The method proposed by Bes and Szymanski improves the Mottelson-Nilsson method by taking into account the short-range pairing correlations:

$$\begin{aligned} E_{\mathrm{BS}} &= \left(\sum_{\nu=1}^{\nu_{\mathrm{F}}} 2e_\nu V_\nu^2 - \frac{\Delta^2}{G} - G \sum_{\nu>0} V_\nu^4 \right)_{\mathrm{p}} \\ &+ \left(\sum_{\nu=1}^{\nu_{\mathrm{F}}} 2e_\nu V_\nu^2 - \frac{\Delta^2}{G} - G \sum_{\nu>0} V_\nu^4 \right)_{\mathrm{n}} + E_{\mathrm{Coul}}. \end{aligned} \quad (10.7)$$

For odd nuclei, a quasiparticle energy has to be added here. These microscopic methods determine nuclear energy with accuracy to a constant. Their absolute value should be finally estimated from the masses of known nuclei and only its variance with deformation can be evaluated from microscopic structure. Nevertherless, the Bes and Szymanski method suffers from the same problem as the Mottelson-Nilson one: double counting of the nucleon-nucleon interaction. The following sections will present a solution of the above disadvantage of the both methods.

10.2 MODERN VERSION OF MACRO-MICRO METHOD

To reproduce absolute values of potential energy and to avoid double summation of nuclear two-body bonds, Strutinsky proposed a method of energy renormalization to liquid drop (or other macroscopic energy) using a microscopic energy correction. The microscopic energy consists of the part responsible for the shell effects and the term associated with pairing correlations. So, the total energy of the nucleus reads

$$E_{\text{Strut}} = E_{\text{LD}} + \delta E_{\text{shell}} + \delta E_{\text{pair}} . \tag{10.8}$$

The contribution of individual terms to Strutinsky's energy (10.8) is schematically presented in Fig. 10.1.

The pairing correction consists of a difference of the microscopic energy calculated with (10.7) and without short-range interactions (10.2) from which the term describing the average energy of the pairing correlations $\langle E_{\text{pair}} \rangle \approx -2.3$ MeV is subtracted, as it has already been included in the phenomenological parameters of the liquid-drop model.

$$\delta E_{\text{pair}} = E_{\text{BCS}} - 2 \sum_{\nu>0}^{\nu_{\text{F}}} e_\nu - \langle E_{\text{pair}} \rangle . \tag{10.9}$$

The shell correction is given by the difference between the sum of the single-particle energies and average energy ($\tilde{E}$) in which the shell structure is washed out.

$$\delta E_{\text{shell}} = 2 \sum_{\nu=1}^{\nu_{\text{F}}} e_\nu - \tilde{E} . \tag{10.10}$$

The idea of separating pure shell effects from nuclear energy, postulated by Myers and Swiatecki in the Fermi gas model, was developed independently by Strutinsky in a more rigorous mathematical model. It will be presented it in the next section. By inserting corrections δE_{pair} (10.9) and δE_{shell} (10.10) to Eq. (10.8), one obtains the equation for the potential energy nuclei in the form

$$E_{\text{Strut}} = E_{\text{LD}} + E_{\text{BCS}} - \langle E_{\text{pair}} \rangle - \tilde{E} . \tag{10.11}$$

The Strutinsky method reproduces the shapes and energies of the nuclei in the ground states and the fission barrier widths and heights. Note that two seemingly opposing nuclear models are included in this method: one based on the assumption of strong interactions with the nuclear macroscopic drop and the second one assuming independent particle moving in an average mean-field potential. These two approaches are not contradictory if one considers the Pauli exclusion principle as discussed in Chapter 3.

10.3 STRUTINSKY SHELL CORRECTION

Strutinsky proposed an efficient way of extracting the so-called smooth part, free from shell structure, from the nucleus energy. In the following, one recalls his argumentation with later improvements made by other authors.

Let us consider a system of A nucleons which coordinates are $\mathbf{x}_i$, $i = 1,\ldots,A$, where $\mathbf{x}_i = \{\mathbf{r}_i,\ \mathbf{s}_i,\ \mathbf{t}_i\}$. The density of i-th nucleon is given by the product $\rho_i(\mathbf{x}_i) = \varphi^\star(\mathbf{x}_i)\cdot\varphi(\mathbf{x}_i)$, where $\varphi(\mathbf{x}_i)$ is its single-particle wave function. The energy of the system consists of a kinetic part, being the sum of the single-particle kinetic energies $T_i(\mathbf{x}_i)$ and potential term originating from the two-body interaction $V(\mathbf{x}_i,\mathbf{x}_j)$

$$
\begin{aligned}
E &= \sum_{i=1}^{A} \int T_i(\mathbf{x}_i)\rho_i(\mathbf{x}_i)\,d\mathbf{x}_i \\
&+ \sum_{\substack{i,j\\ i>j}}^{A} \int\!\!\int \rho_i(\mathbf{x}_i)\,V(\mathbf{x}_i,\mathbf{x}_j)\rho_j(\mathbf{x}_j)\,d\mathbf{x}_i d\mathbf{x}_j\,.
\end{aligned}
\tag{10.12}
$$

The exchange terms (7.24) are omitted here.
Assuming the single-particle mean-field potential in the form

$$
U(\mathbf{x}_i) \equiv \sum_{j\neq i} \int V(\mathbf{x}_i,\mathbf{x}_j)\rho_j(\mathbf{x}_j)d\mathbf{x}_j \tag{10.13}
$$

and taking half of it not to double the nucleon-nucleon interactions, we will express the nucleus energy using the following sum of integrals:

$$
E = \sum_{i=1}^{A} \int \left[T_i(\mathbf{x}_i) + \frac{1}{2}\,U_i(\mathbf{x}_i)\right]\rho_i(\mathbf{x}_i)d\mathbf{x}_i\,. \tag{10.14}
$$

One separates (see Fig. 10.2) the density ρ into a smooth function $\bar{\rho}$ and a fluctuating part $\delta\rho$ containing shell effects

$$
\rho_i(\mathbf{x}_i) = \bar{\rho}_i(\mathbf{x}_i) + \delta\rho_i(\mathbf{x}_i) \tag{10.15}
$$

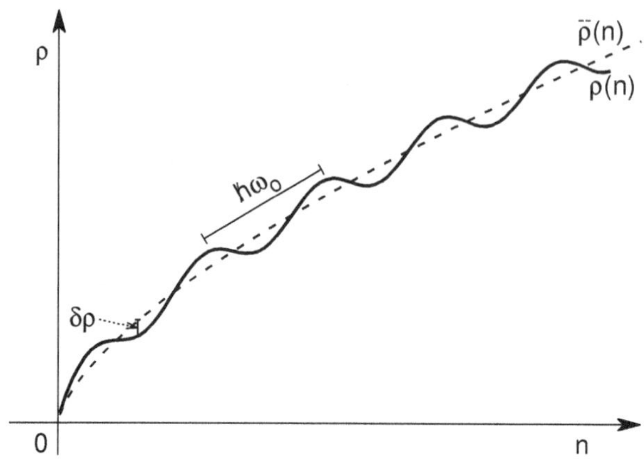

Figure 10.2 Schematic representation of the density of single-particle levels as a function of the number of particles (10.15).

and applies the smooth density $\bar{\rho}$ to Eq. (10.12) to evaluate a nonfluctuating part of energy

$$\begin{aligned}\tilde{E} &= \sum_{i=1}^{A}\int T_i(\mathbf{x}_i)\bar{\rho}_i(\mathbf{x}_i)\,d\mathbf{x}_i + \\ &+ \sum_{i,j,i>j}^{A}\int \bar{\rho}_i(\mathbf{x}_i)\,V(\mathbf{x}_i,\mathbf{x}_j)\bar{\rho}_j(\mathbf{x}_j)d\mathbf{x}_i d\mathbf{x}_j\,.\end{aligned} \tag{10.16}$$

The formula for the total energy of the nucleus can now be written in the form

$$\begin{aligned}E &= \tilde{E} + \left\{\sum_{i=1}^{A}\int T_i(\mathbf{x}_i) + \sum_{j\neq i}^{A}\int V(\mathbf{x}_i,\mathbf{x}_j)\bar{\rho}_j(\mathbf{x}_j)d\mathbf{x}_j\right\}\delta\rho_i(\mathbf{x}_i)d\mathbf{x}_i \\ &+ \text{ terms of higher powers of } \delta\rho\,.\end{aligned} \tag{10.17}$$

By introducing a smooth potential field

$$\tilde{U}(\mathbf{x}_i) = \sum_{j\neq i}^{A}\int V(\mathbf{x}_i,\mathbf{x}_j)\bar{\rho}_j(\mathbf{x}_j)\,d\mathbf{x}_j\,, \tag{10.18}$$

one can convert the formula (10.12) for the total energy of the nucleus to form containing shell correction (10.10)

$$E = \tilde{E} + \sum_{i=1}^{A}\int \left[T_i(\mathbf{x}_i) + \tilde{U}(\mathbf{x}_i)\right]\left[\rho_i(\mathbf{x}_i) - \bar{\rho}_i(\mathbf{x}_i)\right]\,d\mathbf{x}_i \equiv E_{\rm LD} + \delta E_{\rm shell}\,. \tag{10.19}$$

Having deformed average potential $\tilde{U}(\text{def}_\text{U})$, one usually assumes that the surface of the nucleus coincides with the equipotential surface and the potential deformation def_U as well as the density deformation def_ρ are the same. However, this does not have to be always the case. So, in order to improve the quality of the model, one can use def_U to calculate the microscopic correction

$$\text{def}_\text{U} \to U(\text{def}_\text{U}) \to e_\nu \to \delta E_\text{shell}(\text{def}_\text{U}), \tag{10.20}$$

and def_ρ to calculate macroscopic energy

$$\text{def}_\rho \to R(\text{def}_\rho) \to E_\text{LD}(\text{def}_\rho). \tag{10.21}$$

In the majority of calculations, one assumes simply that $\text{def}_\text{U} = \text{def}_\rho$. Such approximation is well justified when one uses, e.g., the Yukawa-folded potential (confer Sec. 6.4), which is generated from a given deformed density distribution. In other cases, one can determine the deformation of the density from a given deformation mean-field potential. Namely, it is possible to evaluate the density deformation of def_ρ assuming the equality of the multipole moments of density distributions calculated in the microscopic and macroscopic way:

$$\text{def}_\text{U} \to U(\text{def}_\text{U}) \to \rho(\text{def}_\text{U}) \to Q^\text{micr}_{\lambda\mu}(\text{def}_\text{U}) = Q^\text{macr}_{\lambda\mu}(\text{def}_\rho) \to \text{def}_\rho. \tag{10.22}$$

Such simplified consistency conditions can improve estimates obtained with the Nilsson or Wood-Saxon potentials.

The question arises now: how to calculate the smoothed sum of single-particle energies? Strutinsky suggested to present the smoothed energy in the form of the following integral over single-particle energies:

$$\tilde{E} = 2\int_{-\infty}^{\tilde{\lambda}} \bar{\rho}(e)e\cdot de. \tag{10.23}$$

The upper limit of the integral, i.e., the Fermi level $\tilde{\lambda}$ in the nucleus with a smoothed shell structure is determined from the conservation condition of a number of particles

$$Z(N) = 2\int_{-\infty}^{\tilde{\lambda}} \bar{\rho}(e)de \longrightarrow \tilde{\lambda}. \tag{10.24}$$

The function $\bar{\rho}$ will be obtained by averaging the true density of single-particle levels ρ using a weight function $j(e, e')$

$$\bar{\rho}(e) = \int_{-\infty}^{+\infty} \rho(e')j(e,e')de', \tag{10.25}$$

where the function j must be normalized

$$\int_{-\infty}^{+\infty} j(e, e')\,de' = 1\,. \tag{10.26}$$

The next question is, how to choose the weight function that has to be im-

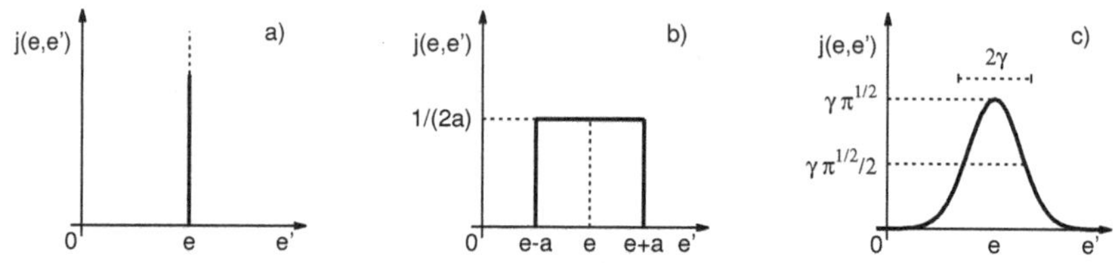

Figure 10.3 Weight functions $j(e, e')$: a) Dirac δ giving strictly localized levels, b) smoothing over a given interval, c) locally smoothing function.

plemented to wash out the single-particle level structure? Let us consider a few examples of it: Dirac δ-function to preserve discrete levels (Fig. 10.3a), smoothing over a given interval (Fig. 10.3b) and a local smoothing function (Fig. 10.3c). The third case is used in practical calculations. Let us discuss each case separately in order to understand better the meaning of the $j(e,e')$ functions shown in Fig. 10.3:

a) For discrete levels, the j weight function would be the Dirac δ-function

$$j(e,e') = \delta(e' - e), \quad \text{where} \quad \int_{-\infty}^{+\infty} \delta(e' - e)\,de' = 1 \tag{10.27}$$

and then

$$\bar{\rho}(e) = \int_{-\infty}^{+\infty} \rho(e')\delta(e - e')\,de' = \rho(e). \tag{10.28}$$

The folding with the Dirac δ-function remains the density unchanged.

b) If the j function is defined as the difference of the Heaviside step functions

$$j(e,e') = \frac{1}{2a}\{\theta[e' - (e-a)] - \theta[e' - (e+a)]\}, \tag{10.29}$$

then one obtains the following smooth density:

$$\begin{aligned}\bar{\rho}(e) &= \frac{1}{2a}\int_{-\infty}^{+\infty}\{\theta[e'-(e-a)]-\theta[e'-(e+a)]\}\rho(e')de' \\ &= \frac{1}{2a}\int_{e-a}^{e+a}\rho(e')de'\,.\end{aligned} \tag{10.30}$$

c) One can also use a modified Gauss function to smooth in energy the spectrum of the single-particle levels

$$j(e,e') = \frac{1}{\gamma\sqrt{\pi}}\,\mathrm{e}^{-(\frac{e-e'}{\gamma})^2} f\left(\frac{e-e'}{\gamma}\right), \tag{10.31}$$

where γ is a smoothing parameter approximately equal to the energy distance between the major shells ($\sim \hbar\omega_0$). The f function is a correction polynomial, ensuring that the average slope of $\bar{\rho}(e)$ is the same as for the $\rho(e)$ curve.

How can the f correction polynomial be determined when weight is the Gauss function?

Let us introduce a variable

$$u = \frac{e-e'}{\gamma}, \tag{10.32}$$

belonging to the range $u \in (-\infty,+\infty)$. The weight function (10.31) can be written as

$$j(e,e') = \frac{\mathrm{e}^{-u^2}}{\gamma\sqrt{\pi}} f(u) \tag{10.33}$$

and the smoothed density function $\bar{\rho}$ depends on the new variable u:

$$\begin{aligned}\bar{\rho}(e') &= \bar{\rho}(e-\gamma u) \equiv \bar{\rho}'(u)\,, \\ \bar{\rho}(e) &= \bar{\rho}(e+\gamma 0) \equiv \bar{\rho}'(0)\,.\end{aligned} \tag{10.34}$$

One can prove that if $\bar{\rho}$ is a polynomial of degree p, then the function f is in the form of a polynomial of the same degree p, and it satisfies the condition

$$\bar{\rho}(e) = \int_{-\infty}^{+\infty}\bar{\rho}(e')\,j(e,e')de'\,, \tag{10.35}$$

ensuring that further averaging should not add anything new to the already smoothed density distribution $\bar{\rho}$.

Let us decompose $\bar{\rho}'(u)$ into a series of Hermite polynomials $H_i(u)$

$$\bar{\rho}'(u) = \sum_{i=1}^{p} a_i H_i(u). \tag{10.36}$$

From the condition (10.35) one receives:

$$\bar{\rho}'(0) = \frac{1}{\sqrt{\pi}} \int_{-\infty}^{+\infty} \bar{\rho}'(u) \mathrm{e}^{-u^2} f(u)\, du \tag{10.37}$$

or after insertion of Eq. (10.36)

$$\sum_{i=1}^{\mathrm{p}} a_i \left\{ \frac{1}{\sqrt{\pi}} \int_{-\infty}^{+\infty} \mathrm{e}^{-u^2} H_i(u) f(u) du - H_i(0) \right\} = 0. \tag{10.38}$$

Assuming that $a_i \neq 0$, the last equation will be satisfied when

$$\frac{1}{\sqrt{\pi}} \int_{-\infty}^{+\infty} \mathrm{e}^{-u^2} H_i(u) f(u)\, du = H_i(0). \tag{10.39}$$

If one also represents the function $f(u)$ in the form of a polynomial of degree r

$$f(u) = \sum_{k=1}^{r} C_k H_k(u), \tag{10.40}$$

then

$$H_i(0) = \sum_{k=1}^{r} C_k \frac{1}{\sqrt{\pi}} \int_{-\infty}^{+\infty} \mathrm{e}^{-u^2} H_i(u) H_k(u)\, du. \tag{10.41}$$

Using the orthogonality properties of the Hermite polynomials, one obtains the last integral equal to : $2^i i! \delta_{ik}$. So,

$$H_i(0) = C_i 2^i i!, \quad \text{hence} \quad C_i = \frac{H_i(0)}{2^i i!}. \tag{10.42}$$

Using the values of the Hermite polynomials at $u = 0$

$$H_i(0) = \begin{cases} 1 & \text{for } i = 0, \\ 2^n \ (-1)^n (2n-1)!! & \text{for } i = 2n, \\ 0 & \text{for } i = 2n+1, \end{cases} \tag{10.43}$$

one obtains

$$C_i = \begin{cases} 1 & \text{for } i = 0, \\ (-1)^n \frac{(2n-1)!!}{2^n (2n)!} & \text{for } i = 2n > 0, \\ 0 & \text{for } i = 2n+1. \end{cases} \tag{10.44}$$

Table 10.1 Expansion coefficients C_i of the correction polynomial (10.40) and the Hermite polynomials $H_i(u)$.

$C_0 = 1$	$H_0 = 1$
$C_2 = -\frac{1}{4}$	$H_2(u) = 4u^2 - 2$
$C_4 = +\frac{1}{32}$	$H_4(u) = 16u^4 - 48u^2 + 12$
$C_6 = -\frac{1}{384}$	$H_6(u) = 64u^6 - 480u^4 + 720u^2 - 120$

Table 10.1 gives the first few expansion coefficients of C_i in the Hermite polynomials series. When evaluating the Strutinsky shell correction energy, one usually uses the correction polynomial of the sixth order, which reads

$$f(u) = \frac{35}{16} - \frac{35}{8}u^2 + \frac{7}{4}u^4 - \frac{1}{6}u^6 . \tag{10.45}$$

Let us summarize the steps in calculating the potential energy of the nucleus by the Strutinsky shell correction method for an even number of particles:

1. First one solves the eigenproblem of a single-particle Hamiltonian $\hat{h}$ with a mean-field potential, which in general is dependent on deformation

$$\hat{h}|\nu\rangle = e_\nu|\nu\rangle \tag{10.46}$$

and one gets the single-particle energy system e_ν as well as eigenfunctions $|\nu\rangle$ labeled with a set of quantum numbers ν.

2. Then, one calculates the smooth part of the single-particle level distribution

$$\bar{\rho}(e) = \frac{1}{\gamma\sqrt{\pi}} \int_{-\infty}^{+\infty} \rho(e') \mathrm{e}^{-\left(\frac{e-e'}{\gamma}\right)^2} f\left(\frac{e-e'}{\gamma}\right) de' , \tag{10.47}$$

where γ is a smoothing constant equal to the mean distances between major shells ($\sim \hbar \overset{\circ}{\omega}_0$).

3. For a given nucleus with Z protons and N neutrons we determine the Fermi level $\tilde{\lambda}$ from the condition

$$\int_{-\infty}^{\tilde{\lambda}} \bar{\rho}(e)de = Z(N). \tag{10.48}$$

4. One calculates the energy of the spectrum of smoothed levels

$$\tilde{E} = 2\int_{-\infty}^{\tilde{\lambda}} e\bar{\rho}(e)de. \tag{10.49}$$

for protons and neutrons.

5. One solves the system of BCS equations and evaluate energy of protons and neutrons in this approximation

$$E_{\text{BCS}} = \sum_{\nu>0} 2e_\nu V_\nu^2 - \frac{\Delta^2}{G} - G\sum_{\nu>0} V_\nu^4. \tag{10.50}$$

6. One calculates the deformation-dependent part of the macroscopic energy, which, e.g., in the liquid-drop model, is (10.1):

$$E_{\text{LD}} = a_2(1-\kappa I^2)A^{2/3}[B_s(\text{def})-1] + \frac{3}{5}\frac{Z^2e^2}{r_0A^{1/3}}[B_c(\text{def})-1]. \tag{10.51}$$

7. One evaluates the total energy of the nucleus as a function of deformation in appropriate grid-points

$$E_{\text{Strut}}(\text{def}) = E_{\text{LD}} + E_{\text{BCS}} - \langle E_{\text{pair}}\rangle - \tilde{E}. \tag{10.52}$$

8. Then one minimizes energy relative to deformation parameters

$$E_{\text{Strut}}(\text{def}) = \text{minimum} \quad\longrightarrow\quad \text{def}_{\text{min}} \tag{10.53}$$

and in this way one can determine all local minima. The lowest one corresponds to the equilibrium (ground-state) point of the nucleus def_{eq}.

9. Having the deformation parameters of the equilibrium point, one can determine the shape of the nucleus $R(\text{def}_{\text{eq}})$ and its ground state energy:

$$E_0 = E_{\text{Strut}}(\text{def}_{\text{eq}}). \tag{10.54}$$

One performs calculations described in points 1 to 5 separately for protons and neutrons, and then the obtained energies $\tilde{E}$ and E_{BCS} are added up. The Strutinsky method is used for various single-particle mean potentials.

The Strutinsky method combines microscopic, macroscopic, and collective deformation effects. It is widely used to evaluate the potential energy surfaces of nuclei in multidimensional space of deformation parameters. An example of the contribution of individual terms to the total Strutinsky energy is shown in Fig. 10.1.

CHAPTER 11

Rotation of atomic nuclei

In the excited spectra of even-even deformed nuclei, there are characteristic rotational bands with energies proportional to $I(I+1)$, where I is the total angular momentum (spin) of the rotating nucleus. These states begin at the energy around 100 keV above ground state and are described by the wave function

$$|IMK\rangle = \left[\frac{2I+1}{8\pi^2}\right]^{1/2} D^I_{MK}(\vartheta_i)\Psi_K(\mathbf{x}),$$

where D^I_{MK} is a generalized spherical function dependent on Euler's angles $\vartheta_1, \vartheta_2, \vartheta_3$ and Ψ_K is a wave function describing the nucleus in internal (i.e., connected with the nucleus) coordinates $\mathbf{x}$. K and M are, respectively, projections of angular momentum I on the axis of symmetry of the nucleus and the axis z of the laboratory system. As in the case of classical rotation, the energy of rotational states of even-even nuclei is determined by the formula

$$E^i_{\text{rot}} = \frac{\hbar^2}{2\mathcal{J}} I(I+1), \tag{11.1}$$

where $\mathcal{J}$ means moment of inertia of the nucleus. A good example of such a rotor is ^{170}Hf nucleus, whose lowest excited levels shown in Fig. 11.1 fulfill very well the $I(I+1)$ rule.

The moment of inertia can be calculated using classical models: rigid-body rotation or rotation of a deformed vessel filled with an ideal non-viscous liquid. Like the quantum mechanical theories, the last model, called the hydrodynamical one, gives the moment of inertia equal to zero when the system is spherically symmetric. The quantum mechanical theories of rotating nuclei

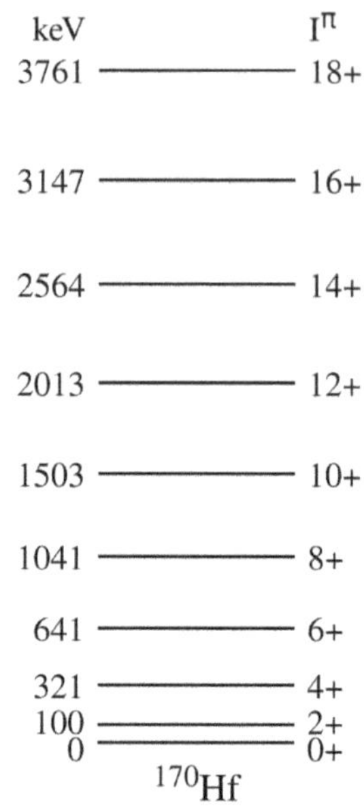

Figure 11.1 Ground state rotational band of ^{170}Hf.

are based on the internal single-particle structure of the nucleus. Namely, the cranking model by Inglis or a model derived in Chapter 15 using the generator coordinate method. In the next three subsections, we will discuss the above models and the deviation of their predictions from a simple rotational rule (11.1).

11.1 RIGID-BODY MODEL

Let us consider the spheroidal, axially symmetric nucleus with the half-axes $a = b$ and c as presented in Fig. 11.2. The axis z is the symmetry axis of the nucleus, and the rotation occurs around axis x. The following equation describes the shape of the rotating nucleus

$$\frac{x^2}{a^2} + \frac{y^2}{b^2} + \frac{z^2}{c^2} = 1 . \tag{11.2}$$

The following formula gives the classic moment of inertia of a rigid body:

$$\mathcal{J}_x^{\rm rig} = \int_{\mathcal{V}} (y^2 + z^2)\rho d^3r , \tag{11.3}$$

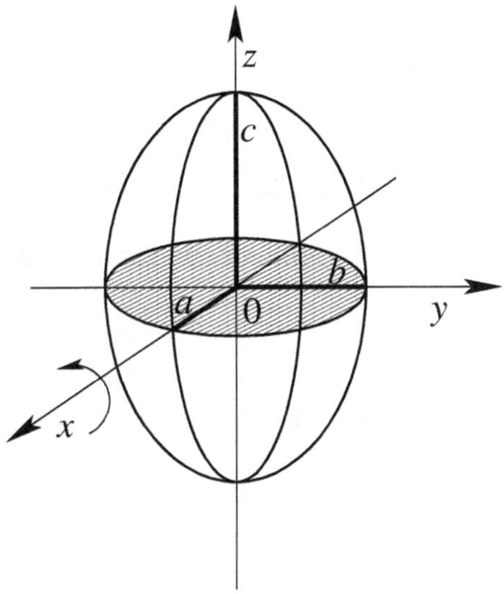

Figure 11.2 Rotation of a rigid spheroid.

where

$$\rho = \frac{M}{\frac{4}{3}\pi a^2 c} \tag{11.4}$$

is density and M mass of the nucleus.
After performing the volume integration, one receives

$$\mathcal{J}_x^{\text{rig}} = \frac{4}{15}\rho a^2 c(a^2 + c^2) = \frac{M}{5}(a^2 + c^2). \tag{11.5}$$

The moment of inertia $\mathcal{J}_x^{\text{rig}}$ of a spheroidally deformed nucleus can be obtained using, e.g., the parameterization developed in the Nilsson model, in which the half-axis lengths are inverse-proportional to the frequencies of the anisotropic oscillator (6.53). The volume conservation condition allows us to express the half-axes a and c using the quadrupole deformation parameters ε (6.43)

$$a = b = R_0 \frac{\overset{\circ}{\omega}_0}{\omega_\perp} = R_0(1 + \frac{\varepsilon}{3})^{-1/3}\left(1 - \frac{2}{3}\varepsilon\right)^{1/3},$$

$$c = R_0 \frac{\overset{\circ}{\omega}_0}{\omega_z} = R_0\left(1 + \frac{\varepsilon}{3}\right)^{2/3}\left(1 - \frac{2}{3}\varepsilon\right)^{-2/3}, \tag{11.6}$$

where $R_0 = 1.2A^{1/3}$ fm is the spherical nucleus radius with the same volume. Hence, one gets the moment of inertia of the nucleus in the rigid body model:

$$\mathcal{J}_x^{\text{rig}} = \frac{2}{5}MR_0^2\left(1 - \frac{1}{3}\varepsilon + \frac{5}{18}\varepsilon^2\right) / \left[\left(1 + \frac{1}{3}\varepsilon\right)^{2/3}\left(1 - \frac{2}{3}\varepsilon\right)^{4/3}\right]. \tag{11.7}$$

The integral (11.3) has to be calculated numerically for nuclear shapes more complex than the ellipsoid.

The rigid-body moment of inertia calculated in the deformed ground state is a few times larger than the experimental one. Another disadvantage of this model is also a non-zero moment of inertia for spherical nuclei ($\varepsilon = 0$), while quantum mechanics demands that $\mathcal{J}(\varepsilon = 0) = 0$ because the Hamiltonian of spherical nucleus commutates with the angular momentum operators. This disadvantage is no longer present in another classical model based on irrotational fluid dynamics, which will be discussed in the next section.

11.2 HYDRODYNAMICAL MODEL

In this model, one assumes the nucleus consists of a non-viscous, incompressible, and irrotational liquid enclosed in a rotating deformed vessel. One calculates the kinetic energy of such nuclear liquid using the classical formula

$$T = \frac{1}{2}\int_{\mathcal{V}} \rho \mathbf{v}_l^2 \, d\tau, \tag{11.8}$$

where $\mathbf{v}_l$ is the velocity of liquid element $d\tau$, ρ the liquid density, and $\mathcal{V}$ the volume of nucleus. Comparing this energy to the formula for rotational energy $T = \mathcal{J}\omega^2/2$ one can determine the moment of inertia

$$\mathcal{J} = \frac{2T}{\omega^2}. \tag{11.9}$$

Let us assume that the rotation of the vessel takes place around the x-axis with angular velocity $\omega = (\omega, 0, 0)$. For an ideal liquid (irrotational and incompressible), one can introduce the scalar potential φ of the velocity field, which fulfills the Laplace equation

$$\hat{\Delta}\varphi = 0.$$

We choose the special solution for this equation in the form

$$\varphi = \mathcal{A}yz.$$

The linear velocity of liquid elements $d\tau$ is determined by the gradient of the scalar potential

$$\mathbf{v}_\mathrm{l} = -\mathbf{grad}\,\varphi = -\frac{\partial\varphi}{\partial x}\mathbf{i} - \frac{\partial\varphi}{\partial y}\mathbf{j} - \frac{\partial\varphi}{\partial z}\mathbf{k} = -\mathcal{A}z\cdot\mathbf{j} - \mathcal{A}y\cdot\mathbf{k}. \tag{11.10}$$

Here **i**, **j**, **k** are the unit vectors along the laboratory axis. To get the constant $\mathcal{A}$, one has to compare the normal component v_l^n of liquid velocity to normal component v_v^n of rotating vessel $\mathbf{v}_\mathrm{v}$ velocity, expressed by

$$\mathbf{v}_\mathrm{v} = \omega\times\mathbf{r} = 0\cdot\mathbf{i} - \omega z\cdot\mathbf{j} + \omega y\cdot\mathbf{k}.$$

The **n** vector is needed to calculate normal components obtained from the ellipsoid equation. To do it, let us consider the equation of the spheroidal vessel

$$f = \frac{x^2}{a^2} + \frac{y^2}{b^2} + \frac{z^2}{c^2} = 1. \tag{11.11}$$

Note that $df = 0$ at the surface, i.e. $(\mathbf{grad}\,f)\cdot d\mathbf{r} = 0$, hence

$$\mathbf{n} = \frac{\mathbf{grad}\,f}{|\mathbf{grad}\,f|} = \frac{2}{|\mathbf{grad}\,f|}\left(\frac{x}{a^2}\mathbf{i} + \frac{y}{b^2}\mathbf{j} + \frac{z}{c^2}\mathbf{k}\right). \tag{11.12}$$

The normal component of the velocity of the rotating vessel is:

$$v_\mathrm{v}^n = \mathbf{v}_\mathrm{v}\cdot\mathbf{n} = (-\omega z\mathbf{j} + \omega y'\mathbf{k})\cdot\left(\frac{x}{a^2}\mathbf{i} + \frac{y}{b^2}\cdot\mathbf{j} + \frac{z}{c^2}\cdot\mathbf{k}\right)\frac{2}{|\mathbf{grad}\,f|} \tag{11.13}$$

and the liquid velocity

$$v_\mathrm{l}^n = \mathbf{v}_\mathrm{l}\cdot\mathbf{n} = (-\mathcal{A}z\mathbf{j} + \mathcal{A}y\mathbf{k})\cdot\left(\frac{x}{a^2}\mathbf{i} + \frac{y}{b^2}\mathbf{j} + \frac{z}{c^2}\mathbf{k}\right)\frac{2}{|\mathbf{grad}\,f|}. \tag{11.14}$$

From the equality $v_\mathrm{v}^n = v_\mathrm{l}^n$ one gets

$$(\mathcal{A}z - \omega z)\frac{y}{b^2} + (\mathcal{A}y + \omega y)\frac{z}{c^2} = 0, \tag{11.15}$$

which gives the expression for the constant $\mathcal{A}$

$$\mathcal{A} = \omega\frac{c^2 - b^2}{c^2 + b^2}. \tag{11.16}$$

Inserting this constant into Eq. (11.10) one determines liquid velocity

$$\mathbf{v}_\mathrm{l} = -\omega\frac{c^2 - b^2}{c^2 + b^2}[z\mathbf{j} - y\mathbf{k}]. \tag{11.17}$$

Having the velocity field, one can already evaluate the kinetic energy (11.8) of liquid filling the rotating spheroid with half axes $a = b$ and c

$$T = \frac{1}{2} \int_{\mathcal{V}} \rho \mathbf{v}_1^2 d^3r, = \frac{3M}{8\pi a^2 c} \left[\frac{(c^2 - a^2)}{(c^2 + a^2)} \omega \right]^2 \int_{\mathcal{V}} (z^2 + y^2) d^3 r.$$

After performing the integration, one obtains

$$T = \frac{M\omega^2}{10} \frac{(c^2 - a^2)^2}{(c^2 + a^2)}. \tag{11.18}$$

Hence, using Eq. (11.9) one gets the moment of inertia in the hydrodynamical irrotational flow model:

$$\mathcal{J}_x^{\mathrm{irr}} = \frac{2}{5} M \frac{c^2 - a^2}{c^2 + a^2}. \tag{11.19}$$

In the Nilsson deformation parameterization ε (6.43) the moment of inertia reads

$$\mathcal{J}_x^{\mathrm{irr}} = \frac{2}{5} M R_0^2 \frac{\varepsilon^2 (1 - \frac{1}{6}\varepsilon)^2}{(1 + \frac{\varepsilon}{3})^{2/3} (1 - \frac{2}{3}\varepsilon)^{4/3} (1 - \frac{\varepsilon}{3} + \frac{5}{18}\varepsilon^2)}. \tag{11.20}$$

Note that for $\varepsilon = 0$, the irrotational flow moment of inertia equals zero. However, $\mathcal{J}_x^{\mathrm{irr}}$ evaluated in the ground state of deformed nuclei is a few times smaller than the experimental moment of inertia.

11.3 CRANKING MODEL

In the Inglis cranking model, one evaluates in the 2nd order perturbation theory the change of nuclear energy when one crank (i.e. rotates) the deformed mean-field potential well around an axis x', which is not the symmetry axis. Let (x', y', z') be the body-fixed coordinates and z' the symmetry axis of the nucleus. Let us assume that rotation occurs with frequency $\omega = (\omega, 0, 0)$ and that the single-particle potential $V(x', y', z')$ does not depend on the time. Of course, it is not the case in the laboratory coordinates (x, y, z) when the potential well rotates.

One has now been reminded that in quantum mechanics, the operator $\hat{R} = \mathrm{e}^{-\mathrm{i}\omega \hat{j}_{x'} t}$ generates the rotations around the x'-axis. Here the operator $\hat{j}_{x'}$ is in $\hbar$ units. The time-dependent Schroedinger equation corresponding to such rotation has the following form in the laboratory frame:

$$\mathrm{i}\hbar \frac{\partial}{\partial t} \psi(x, y, z, t) = \hat{H} \psi(x, y, z, t) \tag{11.21}$$

and it can be transformed to the body-fixed frame using the $\hat{R}$ operator:

$$\psi(x,y,z) = \hat{R}\psi'(x',y',z') \quad \text{and} \quad \hat{H} = \hat{R}\hat{H}'\hat{R}^{+}.$$

After inserting the above relations in Eq. (11.21) one obtains

$$-\mathrm{i}\hbar(\mathrm{i}\omega\hat{j}_{x'})\mathrm{e}^{-\mathrm{i}\omega\hat{j}_{x'}t}\psi' + \mathrm{i}\hbar\mathrm{e}^{-\mathrm{i}\omega\hat{j}_{x'}t}\frac{\partial\psi'}{\partial t} = \mathrm{e}^{-\mathrm{i}\omega\hat{j}_{x'}t}\hat{H}'\mathrm{e}^{+\mathrm{i}\omega\hat{j}_{x'}t}\mathrm{e}^{-\mathrm{i}\omega\hat{j}_{x'}t}\psi' \quad (11.22)$$

and the Schroedinger equation in the body-fixed frame takes the following form:

$$i\hbar\frac{\partial}{\partial t}\psi'(x',y',z',t) = (\hat{H}' - \hbar\omega\hat{j}_{x'})\psi'(x',y',z',t). \quad (11.23)$$

Note that the Routian operator $\hat{\mathcal{R}} \equiv \hat{H}' - \hbar\omega\hat{j}_{x'}$ in the above expression is time-independent. Taking $\hbar\omega\hat{j}_{x'}$ as a small perturbation, one can expand the energy of a rotating system in powers of ω

$$E = E^{(0)} + E^{(1)}\omega + E^{(2)}\omega^2 + \ldots. \quad (11.24)$$

Here, $E^{(0)}$ is the ground-state energy and two other terms one can evaluate using the second-order perturbation theory

$$E^{(1)} = 0, \quad E^{(2)} = \hbar^2\omega^2\sum_{k\neq 0}\frac{|\langle\Phi_k|\hat{j}_{x'}|\Phi_0\rangle|^2}{(E_k - E_0)}, \quad (11.25)$$

where Φ_k is the eigenfunction of the Hamiltonian $\hat{H}'$ in the rotating system, Φ_0 is the ground-state function, and E_0 and E_k are the eigenvalues of $\hat{H}'$. From the above equation and the classical relation $E_{\mathbf{rot}} = \mathcal{J}_x\omega^2/2 = E^{(2)}$ one can determine the cranking moment of inertia

$$\mathcal{J}_x^{\mathrm{cr}} = 2\hbar^2\sum_{k\neq 0}\frac{|\langle\Phi_k|\hat{j}_{x'}|\Phi_0\rangle|^2}{(E_k - E_0)}. \quad (11.26)$$

In the absence of the pairing correlations, this microscopic formula gives the values of the moment of inertia close to the hydrodynamical model estimates for near magic nuclei, while for well deformed nuclei, its estimates are not far from the rigid-body ones, i.e., they are relatively far from the experimental data.

To determine moments of inertia in a superconducting model of nuclei, let us write the $\hat{j}_x$ operator in the second quantization language:

$$\hat{j}_x^{\mathrm{II}} = \sum_{\mu\nu}\langle\mu|\hat{j}_{x'}|\nu\rangle a_\nu^{+}a_\mu, \quad (11.27)$$

where μ, ν are the single-particle functions.

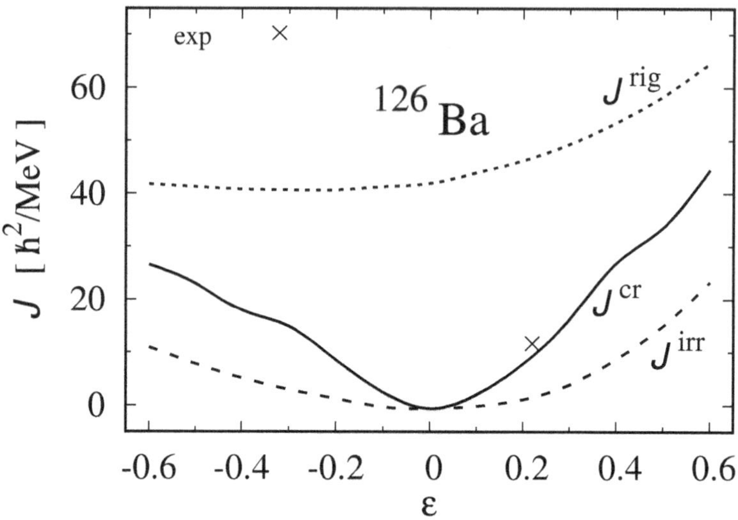

Figure 11.3 Moment of inertia of ^{126}Ba calculated according to the rigid-body $J^{\rm rig}$, cranking $J^{\rm cr}$ and irrotational flow $J^{\rm irr}$ models as a function of the quadrupole deformation ε. The experimental value of $\mathcal{J}$ is marked by a cross at the equilibrium deformation $\varepsilon_{\rm eq} = 0.22$.

We assume also that the Hamiltonian $\hat{H}'$ contains the pairing correlations and its excited states are two quasiparticle ones for even-even nuclei with energies $E_\nu + E_\mu$, and eigenfunctions being two-quasiparticle states $|\Phi_k\rangle = \alpha_\nu^+ \alpha_\mu^+ |\Phi_0\rangle$, created on the BCS ground state $|\Phi_0\rangle$ function (9.3). After transforming the operator $\hat{j}_x$ (11.27) into the quasiparticles (9.16) one can easily evaluate the matrix element in Eq. (11.26) and one obtains the formula for the moment of inertia

$$\mathcal{J}_x^{\rm cr} = 2\hbar^2 \sum_{\mu\nu} \frac{|\langle\mu|\hat{j}_{x'}|\nu\rangle|^2 (U_\nu V_\mu - V_\nu U_\mu)^2}{E_\nu + E_\mu}, \tag{11.28}$$

where $E_\nu = \sqrt{(e_\nu - \lambda)^2 + \Delta^2}$ is the quasiparticle energy and U_ν, V_μ are the BCS occupation amplitudes (9.27).

Fig. 11.3 shows the dependence on quadrupole deformation ε of the moments of inertia calculated for ^{126}Ba according to the three above models. The moments of inertia obtained in the rigid-body model $\mathcal{J}^{\rm rig}$ are greater than those for the irrotational flow approximation $\mathcal{J}^{\rm irr}$ and the cranking theory $\mathcal{J}^{\rm cr}$. The value of the moment of inertia $\mathcal{J}^{\rm cr}$ ^{126}Ba calculated in equilibrium deformation $\varepsilon^{eq} = 0.22$ is close to the experimental data (cross).

11.4 DEVIATIONS FROM THE ROTATION MODEL

The main assumptions of the rotational model are:

- proportionality of excited level energies to the square angular momentum: $E^i_{\text{rot}} \sim I(I+1)$,
- unchanged deformation of the rotating nucleus.

The second point can be experimentally tested by looking at the reduced transition probabilities between the states from the rotational band

$$B(E\lambda, I_\text{i} \to I_\text{f}) \sim Q^2_\lambda. \tag{11.29}$$

The data show that the values of $B(E2, I \to I-2)$ slightly change with growing I, which tells that the quadrupole moment Q_2, i.e., quadrupole deformation, is not constant. It also turns out that for larger spins, the I^2 rule is not sufficient, and one has to add higher powers of I to reproduce the energies of the rotational states

$$E_{\text{rot}}(I) = \frac{\hbar^2}{2\mathcal{J}} I(I+1) + BI^2(I+1)^2 + \dots . \tag{11.30}$$

The above deviations from the rotational model may have the following origin:

- contracting or stretching a rotating nucleus — deformation and moment of inertia $\mathcal{J}$ of nucleus usually grows with increasing spin,
- change of quasiparticle configurations in the rotating nuclei,
- reduction or even breaking of the pairing correlations in a fast-rotating nucleus due to Coriolis forces, which decouple pairs of nucleons conjugated to $J = 0$. This effect makes the rotational levels more compact.

Phenomena accompanying faster and faster rotation of the nucleus can be observed on the *yrast line*[1], which joins the states having the minimal energy a given angular momentum I. In Fig. 11.4, the energies of excited states (points) of ^{152}Dy are plotted as a function of their spin. The yrast line joins the states with the lowest energy; above it, there is an area in which various excited states can occur. It is visible that some of levels form a kind of band. By definition, there are no levels below the yrast line , while one can find the statistically distributed levels at much higher energies than the yrast line.

[1]The word yrast comes from Swedish and means fast rotation.

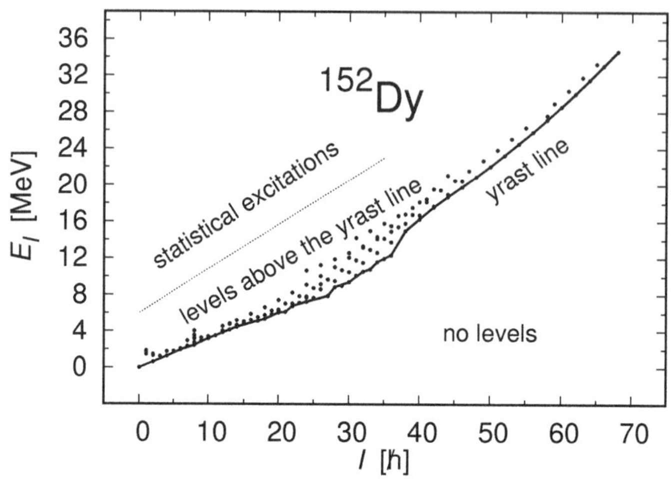

Figure 11.4 Energies of excited states of ^{152}Dy as a function of spin I (points). The solid line, called the yrast line, joins the states having the lowest energy for a given angular momentum I.

In well-deformed nuclei, the rotational energy levels with $I \leq 10\hbar$ are described by formula (11.1), but this simple rule does not work well for higher spins. When angular momentum increases, one even observes a jump of the local (called *dynamic*) moment of inertia $\mathcal{J}^{(2)} = d^2E/d^2I$ as a function of the rotation frequency $\omega = dE/dI$ as it presented in Fig. 11.5 for ^{158}Er. Such a jump is called the *back-bending* phenomenon. It is related to the rapid change of the quasiparticle configuration: a rotational band built on a two-quasiparticle configuration (S) crosses the ground-state band. The pairing correlation in the two-quasiparticle states is weaker (see Sec. 9.5), so the corresponding moment of inertia is larger than in the ground state band. Consequently, its angular velocity should decrease to have the same angular momentum in both bands. One can also show that the centrifugal stretching in fast rotating nuclei leads first to the oblate forms (larger moment of inertia), then through non-axial shapes to very elongated prolate forms of the nucleus, and finally, at a critical angular momentum value I_{crit}, even to splitting of a nucleus into two fragments.

Some rotational bands above the yrast line may be related to the shape isomers, i.e., local minima in the potential energy surface corresponding to very deformed shapes. Particularly strong shell effects occur in the elongated shapes corresponding to the ratio of the spheroid half-axes 2:1. Such *superdeformation* was discovered in 1960 in isotopes of fissioning atomic nuclei

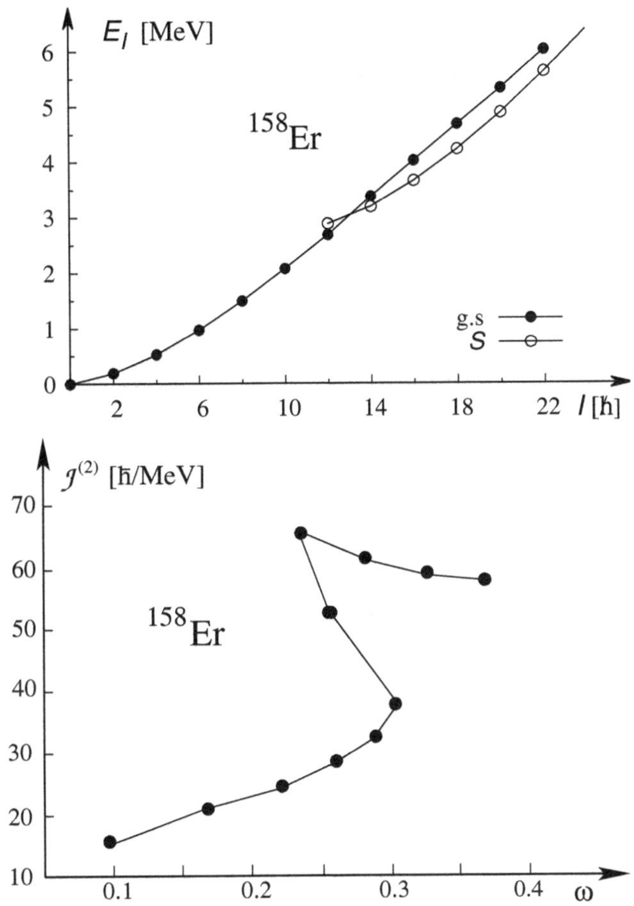

Figure 11.5 Ground state (g.s.) and S rotational bands (top.) of ^{158}Er and the corresponding dynamic moment of inertia $\mathcal{J}^{(2)}$ as a function of the rotational frequency ω. The back-bending phenomenon (bottom) appears here at the band crossing around spin $I = 14$.

(e.g., ^{240}Pu), and 1980 in fast-rotating atomic nuclei (e.g., ^{152}Dy). In addition to super-deformation, hyper-deformation corresponding to shapes with a half-axises ratio around 3:1 is observed in some Th isotopes.

To calculate the energies of rotational states at high spins, one has to solve exactly Eq. (11.23) but not by a perturbation calculation like it was done in the cranking model. The BCS formalism can also be applied here but with an additional constraint for I. One has to solve the variational problem for the

Hamiltonian with the pairing term

$$\langle\Phi|\hat{H}|\Phi\rangle = \text{minimum} \tag{11.31}$$

with conditions to preserve the number of particles

$$\langle\Phi|\hat{N}|\Phi\rangle = n \tag{11.32}$$

and angular momentum

$$\langle\Phi|\hat{I}|\Phi\rangle = I\,. \tag{11.33}$$

So, one is looking for the minimum of functional

$$\langle\Phi|\hat{H} - \lambda\hat{N} - \omega\hat{I}|\Phi\rangle = \text{minimum}\,, \tag{11.34}$$

where λ and ω are Lagrange multipliers.

In the following, one considers this problem in the model case only when 'rotation' [2] takes place around the symmetry axis. It is not a purely academic problem as such a type of 'rotation' is preferred by oblate nuclei. Let us assume the 'rotation' around the z axis, which is the symmetry axis of the nucleus. The corresponding angular momentum operator reads

$$\hat{I} = \hat{I}_z = \sum_\nu m_\nu a_\nu^+ a_\nu\,, \tag{11.35}$$

where m_ν is the third component of the angular momentum in the state $|\nu\rangle$: $m_\nu = \langle\nu|\hat{I}_z|\nu\rangle$.

Transforming the operator $\hat{I}_z$ to the quasiparticle image and using the formula (9.18) one can rewrite Eq. (11.34) in the form

$$\begin{aligned}\langle\Phi|\hat{H} - \lambda\hat{N} - \omega\hat{I}_z|\Phi\rangle \;&= 2\sum_\nu (e_\nu - \lambda - \omega m_\nu)V_\nu^2 \\ &-G\left(\sum_\nu U_\nu V_\nu\right)^2 - G\sum_\nu V_\nu^4 = \text{minimum}\,.\end{aligned} \tag{11.36}$$

From here (see Sec. 9.1) the filling probabilities of the state $|\nu\rangle$ are

$$V_\nu^2 = \frac{1}{2}\left(1 - \frac{e_\nu - \lambda - \omega m_\nu - GV_\nu^2}{\sqrt{(e_\nu - \lambda - \omega m_\nu - GV_\nu^2)^2 + \Delta^2}}\right), \tag{11.37}$$

and $U_\nu^2 = 1 - V_\nu^2$.

[2]The quotation marks have been used here, as in this case, one can speak only about a classic analogy of rotation because, in quantum mechanics, rotation around the symmetry axis does not exist.

Note that the single-particle states $|\nu\rangle$ are no longer twice degenerated in energies due to rotation. Of course, besides Eq. (11.36), two additional constraints for particle number and angular momentum must be fulfilled:

$$\sum_{\nu} V_{\nu}^{2} = n\,, \tag{11.38}$$

$$\sum_{\nu} V_{\nu}^{2} m_{\nu} = I\,. \tag{11.39}$$

In practice, one calculates the total energy for a rotating nucleus for different values of ω on a grid of deformation points $E_{\text{Strut}}^{\text{rot}}(\text{def};\omega)$ in an analogous way to that described Eqs. (10.46-10.54). After minimizing energy with respect to deformation one gets energy $E_0^{\text{rot}}(\omega)$ and angular momentum $I(\omega)$ of the nucleus at the equilibrium point, whose location depends on ω. With this data, one can construct the yrast line.

The appearance of the $-\omega m_\nu$ term in single-particle energies in the formula for V_ν^2 causes the Fermi level to be tilted. A graphic illustration of this effect is shown in Fig. 11.6. On the plane $(e_\nu, 2m)$, the states filled with particles are marked by black points and unoccupied with empty circles.

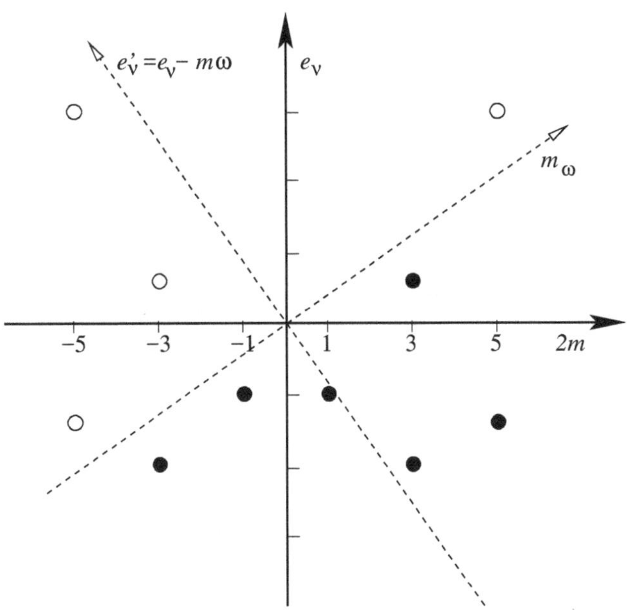

Figure 11.6 Filling of single-particle states e_ν below the tilted Fermi level (dashed line, m_ω) whose slope depends on angular velocity ω.

In the case of collective rotations of the nuclei, i.e., not around the axis of symmetry, the eigenproblem with pairing forces is solved in a more complex way than in the model case discussed above, but the idea of the calculation is the same.

Spectroscopy of high-spin states of atomic nuclei is currently the subject of extensive theoretical and experimental research, and a more detailed discussion of the results of these works would go beyond the limits of this book.

CHAPTER 12

Collective vibrations of nuclei

Collective excitation modes in nuclei have been well-known and studied since the early 1950s. They were observed in photo-absorption experiments, with a probability two orders of magnitude larger than the single-particle excitation.

Contrary to the low-lying rotational states, the first excited vibrational states of even-even atomic nuclei have a higher energy of around 1 MeV, lower than the typical two-quasi-particle energy of about 2 MeV. They are related to collective oscillations of nuclei. Depending on their origin, one can speak about:

- shape vibration,
- vibrations of radial density distribution,
- collective pairing vibration.

All the above types of nuclear excitations are related to oscillations of the mean-field.

In addition to these low-energy vibrations, there are collective excitations of much higher energies of up to a dozen or more MeV. Those are:

- shift of the proton density with respect to the density of the neutron, e.g., giant dipole resonance,
- density oscillations – waves in nuclear matter.

All these types of vibrations can be described by the macroscopic method using liquid-drop-like models or microscopically, e.g., by the Hartree-Fock method using an adiabatic approximation in which one assumes that the collective vibrations are slower than rapid single-particle motion.

12.1 MACROSCOPIC DESCRIPTION OF VIBRATIONS

Let us expand the deviation of the nuclear surface from a sphere in a series of spherical harmonics like it was described in Eq. (5.44), and assume that the expansion coefficients $\alpha_{\lambda\mu}(t)$ may change in time. For small surface oscillations around the sphere, the energy E_λ corresponding to the change of deformation of multipolarity λ can be approximated by

$$E_\lambda = \frac{C_\lambda}{2} \sum_\mu |\alpha_{\lambda\mu}|^2 + \frac{B_\lambda}{2} \sum_\mu |\dot{\alpha}_{\lambda\mu},|^2 \tag{12.1}$$

where C_λ is the stiffness parameter related to the change of surface and Coulomb energies with deformation of multipolarity λ. The coefficients B_λ in front of the square of velocities $\dot{\alpha}_{\lambda\mu}$ play a role of the collective mass parameters (inertia). Replacing the classic oscillator energy (12.1) by the quantum Hamiltonian (see Sec. 6.1), one can get an equidistant spectrum with a distance between levels corresponding to harmonic oscillator frequency of multipolarity λ

$$\hbar\omega_\lambda = \hbar \sqrt{\frac{C_\lambda}{B_\lambda}}. \tag{12.2}$$

For a harmonic oscillator (see quantum mechanics textbooks), the reduced probability $B(\mathrm{E}\lambda)$ for the gamma transition of multipolarity λ between states of rotational bands built up on neighboring vibrational states is given by the following formula

$$B(\mathrm{E}_\lambda) = (2\lambda + 1) \left(\frac{3ZeR_0^\lambda}{4\pi} \right)^2 \frac{\hbar}{2B_\lambda \omega_\lambda}. \tag{12.3}$$

Knowing the distance between the neighboring vibrational states $\hbar\omega_\lambda$ and the reduced transition probability one can estimate, using the two above equations, the stiffness and inertia parameters. The B_λ parameters can also be estimated in the hydrodynamical model (see Sec. 12.2), while C_λ is calculated macroscopically within the nuclear liquid drop model. As one can expect, the macroscopic estimates of both parameters do not agree with the parameters extracted from the experimental data. It is unsurprising as a similar problem was met when discussing the rotational spectra. So, it is better to evaluate B_λ and C_λ using a model that considers the nucleus's single-particle structure, i.e., by calculating them microscopically.

12.2 MICROSCOPIC DESCRIPTION OF VIBRATIONS

Among theoretical methods describing the vibrations of the nucleus microscopically, one can distinguish the random phase approximation (RPA), the interacting boson model (IBM), the cranking method, the time-dependent Hartree-Fock approximation, and the generator coordinate method. Below, the mean features of the above theories will be presented, apart from the IBM method, which demands from the reader a good knowledge of the group theory.

12.2.1 Dispersion equation

Using the time-dependent Hartree-Fock method, one can derive a dispersion equation in which the solution gives the energy of the lowest vibrational state. The eigenfunctions of a single-particle Hamiltonian with a given mean-field potential $\hat{h} = \hat{T} + V(\text{def})$ will be used as a basis to represent the many-body Hamiltonian $\hat{H}$ in the second quantization language. One also assumes that $\hat{H}$ contains a pairing term responsible for the short-range (9.1) correlations. An approximate solution of such a Hamiltonian eigenproblem can be obtained using, e.g., the BCS theory and the quasiparticle representation

$$\hat{H}|\Phi_0\rangle = E_0|\Phi_0\rangle, \qquad \hat{H}|\Phi_j\rangle = E_j|\Phi_j\rangle, \tag{12.4}$$

where $|\Phi_0\rangle$ is the BCS ground state function (9.3), E_0 is the ground state energy (9.33) and $|\Phi_j\rangle$ are two-particle excited states (9.47) with energy E_j (9.48). The eigenstates of $\hat{H}$ are orthogonal $\langle\Phi_j|\Phi_k\rangle = \delta_{jk}$.

Going towards the time-dependent problem, one constructs a Hamiltonian with a residual, e.g., multipole-multipole $\hat{F}$ interaction, which generates vibration (see definition (9.18))

$$\hat{\mathcal{H}} = \hat{H} - \frac{\kappa}{2}\hat{F}\hat{F}, \tag{12.5}$$

where κ is the strength of the interaction.

According to the Hartree-Fock method, one linearizes $\hat{\mathcal{H}}$ to a form of the following single-particle Hamiltonian:

$$\hat{\mathcal{H}} = \hat{H} - \kappa F\hat{F}. \tag{12.6}$$

Then, one assumes that the average value of the $\hat{F}$ operator oscillates with a frequency ω

$$F = \langle\psi|\hat{F}|\psi\rangle = F_0\cos(\omega t), \tag{12.7}$$

where ψ is the many-body time-dependent wave function of the whole nucleus.

To get self-consistency at any time, one has to solve the eigenproblem of the time-dependent Schroedinger equation with Hamiltonian $\hat{\mathcal{H}}$. Then the average value of $\hat{F}$ between assumed functions $|\psi\rangle$ has to be compared to F, which leads to the so-called dispersion equation for eigenvalues of $\hat{\mathcal{H}}$.

We are going to find the solution of the Schroedinger equation

$$i\hbar\frac{\partial}{\partial t}|\psi\rangle = \hat{\mathcal{H}}|\psi\rangle, \tag{12.8}$$

in a standard form

$$|\psi\rangle = \sum_j c_j(t)|\Phi_j\rangle e^{-\frac{i}{\hbar}E_j t}, \tag{12.9}$$

i.e., by expanding the time-dependent function $|\psi\rangle$ in the eigenfunctions (12.4) of the stationary Hamiltonian $\hat{H}$. For small vibrations, when $\hbar\omega < E_i - E_0$, the coefficients c_j can be evaluated using the perturbation calculation

$$c_j = c_j^0 + c_j' + \dots, \tag{12.10}$$

where $c_j^0 = \delta_{j0}$ and c_j' is small. This allows to separate the ground state from the rest of the sum (12.9)

$$|\psi\rangle = |\Phi_0\rangle e^{-\frac{i}{\hbar}E_0 t} + \sum_{j\neq 0} c_j'|\Phi_j\rangle e^{-\frac{i}{\hbar}E_j t}. \tag{12.11}$$

Inserting the function (12.9) to the time-dependent Schroedinger equation (12.8) one obtains

$$i\hbar\left[\sum_j \dot{c}_j|\Phi_j\rangle e^{-\frac{i}{\hbar}E_j t} - \frac{i}{\hbar}\sum_j c_j E_j|\Phi_j\rangle e^{-\frac{i}{\hbar}E_j t}\right] = \sum_j c_j E_j|\Phi_j\rangle e^{-\frac{i}{\hbar}E_j t}$$
$$-\kappa F\sum_j \hat{F}|\Phi_j\rangle c_j e^{-\frac{i}{\hbar}E_j t}. \tag{12.12}$$

By multiplying the above equation from the left side by $\langle\Phi_k|$ and using orthogonality conditions, one receives

$$i\hbar\dot{c}_k e^{-\frac{i}{\hbar}E_k t} = -\kappa F\sum_j \langle\Phi_k|\hat{F}|\Phi_j\rangle c_j e^{-\frac{i}{\hbar}E_j t}. \tag{12.13}$$

Then, with the assumption (12.10) one obtains its simplified form

$$i\hbar\dot{c}_k' e^{-\frac{i}{\hbar}E_k t} = -\kappa F\left(\sum_{j\neq 0}\langle\Phi_k|\hat{F}|\Phi_j\rangle c_j' e^{-\frac{i}{\hbar}E_j t} + \langle\Phi_k|\hat{F}|\Phi_0\rangle e^{-\frac{i}{\hbar}E_0 t}\right). \tag{12.14}$$

Omitting the quadratic terms in c'_j, i.e., neglecting the matrix elements of the interactions between excited states, one gets:

$$c'_k = \frac{-\kappa}{i\hbar}\langle\Phi_k|\hat{F}|\Phi_0\rangle \int F_0\cos(\omega t)\,e^{\frac{i}{\hbar}(E_k-E_0)t}dt, \qquad (12.15)$$

then, after performing the integration, one obtains

$$c'_k = \frac{\kappa}{2}\langle\Phi_k|\hat{F}|\Phi_0\rangle F_0\left[\frac{e^{\frac{i}{\hbar}[(E_k-E_0)+\hbar\omega]t}}{(E_k-E_0+\hbar\omega)} + \frac{e^{\frac{i}{\hbar}[(E_k-E_0)-\hbar\omega]t}}{(E_k-E_0-\hbar\omega)}\right]. \qquad (12.16)$$

One assumed here $\hbar\omega \ll E_i - E_0$ and that $c'_j(t=0) = 0$, which means that at the beginning, the system was in the ground-state $|\Phi_0\rangle$.

Let us now calculate the average value $\hat{F}$ from the equation (12.7) using the function ψ given by Eq. (12.11)

$$F=\langle\Phi_0|\hat{F}|\Phi_0\rangle+\left[\sum_{j\neq 0} c'_j\langle\Phi_0|\hat{F}|\Phi_j\rangle e^{\frac{-i}{\hbar}(E_j-E_0)t} + \sum_{j\neq 0} c'^{\star}_j\langle\Phi_j|\hat{F}|\Phi_0\rangle e^{\frac{i}{\hbar}(E_j-E_0)t}\right]\ldots . \qquad (12.17)$$

The first term is zero because the average value of long-range interactions has already been included in the deformed Hamiltonian. Similarly, as in Eq. (12.14), one has omitted the terms $\langle\Phi_j|\hat{F}|\Phi_k\rangle \approx 0$ for $j \neq 0$ and $k \neq 0$. Finally, after insertion (12.16), the average value of the $\hat{F}$ the operator takes the form

$$\begin{aligned} F &= \kappa F_0 \sum_{j\neq 0} |\langle\Phi_j|\hat{F}|\Phi_0\rangle|^2 \frac{(e^{i\omega t}+e^{-i\omega t})(E_j-E_0)}{(E_j-E_0-\hbar\omega)(E_j-E_0+\hbar\omega)} = \\ &= 2\kappa F_0\cos(\omega t)\sum_{j\neq 0}\frac{|\langle\Phi_j|\hat{F}|\Phi_0\rangle|^2(E_j-E_0)}{(E_j-E_0)^2-(\hbar\omega)^2}. \end{aligned} \qquad (12.18)$$

Comparing the obtained expression with the assumed in Eq. (12.7) value $F_0\cos\omega t$, one gets the dispersion equation, analogous to the condition for the refractive coefficient in optics

$$\frac{1}{\kappa} = 2\sum_{j\neq 0}\frac{|\langle\Phi_j|\hat{F}|\Phi_0\rangle|^2(E_j-E_0)}{(E_j-E_0)^2-(\hbar\omega)^2} \equiv S(\omega). \qquad (12.19)$$

This equation can be solved graphically, as it is shown schematically in Fig. 12.1. The line corresponding to $S(\omega)$ runs between asymptotes at $(E_i - E_0)$, for which $S(\omega) = \pm\infty$. The intersection of the graph with the horizontal

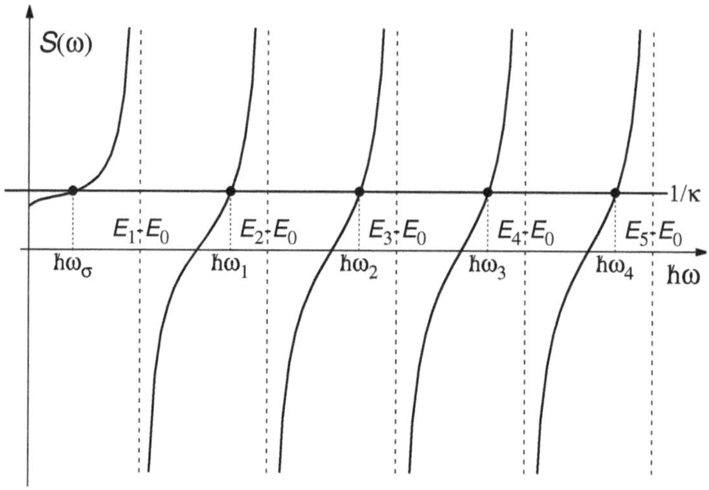

Figure 12.1 Graphic solution of the dispersion equation.

straight line at $1/\kappa$ gives all possible solutions satisfying Eq. (12.19). The first one, $\hbar\omega_\sigma$ is strongly lowered, and it can describe collective states of $\lambda = 0, 2, 3\ldots$ multipolarity, depending on the definition of $\hat{F}$. Note, that the value of $S(\omega)$ grows with the number of base states (if the matrix element $\langle\Phi_j|\hat{F}|\Phi_0\rangle$ does not vanish for $|\Phi_j\rangle$ belonging to higher shells), and for a too-large number of them, it may occur that the sum $S(\omega = 0)$ will be greater than $1/\kappa$. It is a minor point of the above method. The other solutions of Eq. (12.19) are located close to the poles $(E_i - E_0)$ as is shown in Fig. 12.1 and they have less collective nature.

12.2.2 Mass and stiffness parameters

By calculating the energy of the system in the state ψ (12.11) and comparing it with the formula (12.1), one can get expressions for the mass and stiffness parameters.

Let us evaluate the energy in the state $|\psi\rangle$

$$E = \frac{\langle\psi|\hat{\mathcal{H}}|\psi\rangle}{\langle\psi|\psi\rangle} = \frac{\langle\psi|\hat{H}|\psi\rangle}{\langle\psi|\psi\rangle} - \frac{\kappa}{2}\frac{\langle\psi|\hat{F}^2|\psi\rangle}{\langle\psi|\psi\rangle}. \tag{12.20}$$

Using the orthogonality conditions of $|j\rangle$ states, one gets the norm of the state ψ

$$\langle\psi|\psi\rangle = 1 + \sum_j |c_j|^2 \tag{12.21}$$

and the average value of the Hamiltonian $\hat{H}$

$$\langle\psi|\hat{H}|\psi\rangle = E_0 + \sum_j |c_j|^2 E_j . \tag{12.22}$$

In addition, one assumes that

$$\frac{\langle\psi|\hat{F}^2|\psi\rangle}{\langle\psi|\psi\rangle} \approx \left|\frac{\langle\psi|\hat{F}|\psi\rangle}{|\langle\psi|\psi\rangle}\right|^2 = F^2 , \tag{12.23}$$

which is a simple consequence of the Hartree-Fock approximation (12.6). Hence

$$E = \frac{E_0 + \sum_j |c_j|^2 E_j}{1 + \sum_j |c_j|^2} - \frac{\kappa}{2} F^2 , \tag{12.24}$$

and omitting the small term $(\sum_j |c_j|^2)$ in the denominator, one obtains

$$E = E_0 + \sum_j |c_j|^2 (E_j - E_0) - \frac{\kappa}{2} F^2 . \tag{12.25}$$

Using the expression for the expansion coefficients:

$$c_j = \delta_{j0} + \frac{\kappa F_0}{2} \langle\Phi_j|\hat{F}|\Phi_0\rangle \left\{ \frac{e^{\frac{i}{\hbar}(E_j - E_0 + \omega\hbar)t}}{E_j - E_0 + \hbar\omega} + \frac{e^{\frac{i}{\hbar}(E_j - E_0 - \hbar\omega)t}}{E_j - E_0 - \hbar\omega} \right\} , \tag{12.26}$$

one obtains for $j \neq 0$

$$\begin{aligned} |c_j|^2 &= \frac{\kappa^2 F_0^2}{4} |\langle\Phi_j|\hat{F}|\Phi_0\rangle|^2 \\ &\cdot \left[\frac{4\omega^2\hbar^2}{[(E_j - E_0)^2 - (\hbar\omega)^2]^2} + \frac{4\cos^2(\omega t)}{(E_j - E_0)^2 - (\hbar\omega)^2} \right] . \end{aligned} \tag{12.27}$$

Hence, the total energy of the vibrating nucleus (12.25) can be written as

$$\begin{aligned} E &= E_0 + \sum_j \frac{\kappa^2 \hbar^2 \omega^2 F_0^2 |\langle\Phi_j|\hat{F}|\Phi_0\rangle|^2 (E_j - E_0)}{[(E_j - E_0)^2 - (\hbar\omega)^2]^2} \\ &\quad + \kappa^2 F_0^2 \sum_j \langle\Phi_j|\hat{F}|\Phi_0\rangle|^2 \frac{(E_j - E_0)}{(E_j - E_0)^2 - (\hbar\omega)^2} \cos^2(\omega t) \\ &\quad - \frac{\kappa}{2} F_0^2 \cos^2(\omega t) \equiv E_{\mathrm{I}} + E_{\mathrm{II}} \cos^2(\omega t) , \end{aligned} \tag{12.28}$$

where

$$E_{\rm I} \equiv E_0 + \sum_j \frac{\kappa^2\hbar^2\omega^2 F_0^2 |\langle\Phi_j|\hat{F}|\Phi_0\rangle|^2 (E_j - E_0)}{[(E_j - E_0)^2 - (\hbar\omega)^2]^2}, \tag{12.29}$$

$$E_{\rm II} \equiv \frac{\kappa^2 F_0^2}{2}\left\{2\sum_j \frac{|\langle\Phi_j|\hat{F}|\Phi_0\rangle|^2 (E_j - E_0)}{(E_j - E_0)^2 - (\hbar\omega)^2} - \frac{1}{\kappa}\right\}. \tag{12.30}$$

When the dispersion equation (12.19) is fulfilled, then $\omega = \omega_\sigma$, the second term disappears $E_{\rm II} = 0$, and the energy does not depend on the time, i.e., it is stationary

$$E = E_0 + \kappa^2 F_0^2 (\hbar\omega_\sigma)^2 \sum_j \frac{|\langle\Phi_j|\hat{F}|\Phi_0\rangle|^2 (E_j - E_0)}{[(E_j - E_0)^2 - (\hbar\omega_\sigma)^2]}^2. \tag{12.31}$$

Let us present this energy in a different form. Differentiating $F = F_0\cos(\omega t)$ versus time, one gets $\dot{F} = -F_0\omega\sin(\omega t)$, which gives

$$\sin(\omega t) = -\frac{\dot{F}}{F_0\omega} \quad \text{and} \quad \cos(\omega t) = \frac{F}{F_0}. \tag{12.32}$$

Hence taking the relation $\sin^2\omega + \cos^2\omega = 1$ one obtains

$$\dot{F}^2 + F^2\omega^2 = F_0^2\omega^2. \tag{12.33}$$

By inserting the last equality into the formula (12.31), one gets

$$E = E_0 + \kappa^2\hbar^2(\dot{F}^2 + F^2\omega_\sigma^2) \sum_j \frac{|\langle\Phi_j|\hat{F}|\Phi_O\rangle|^2 (E_j - E_0)}{[(E_j - E_0)^2 - (\hbar\omega_\sigma)^2]^2}. \tag{12.34}$$

The above equation expressed as

$$E = E_0 + \frac{1}{2}B_\sigma \dot{F}^2 + \frac{1}{2}C_\sigma F^2, \tag{12.35}$$

leads to the formula for the collective inertia

$$B_\sigma = 2\kappa^2 \sum_j \frac{|\langle\Phi_j|\hat{F}|\Phi_0\rangle|^2 (E_j - E_0)}{[(E_j - E_0)^2 - (\hbar\omega_\sigma)^2]^2} \tag{12.36}$$

and the stiffness parameter

$$C_\sigma = \omega_\sigma^2 B_\sigma. \tag{12.37}$$

ω_σ is therefore the frequency of harmonic oscillation.

12.2.3 Adiabatic approximation

Suppose the frequency of vibrations ω is small compared to the energy difference of excited states, $\hbar\omega \ll E_j - E_0$. In that case, one can get an analytical solution (called adiabatic) of the dispersion equation (12.19). To show it, let use the following identity:

$$\frac{1}{(E_j - E_0)^2 - (\hbar\omega)^2} = \frac{1 + \left(\frac{\hbar\omega}{E_j - E_0}\right)^2}{(E_j - E_0)^2 \left[1 - \left(\frac{\hbar\omega}{E_j - E_0}\right)^4\right]}, \tag{12.38}$$

then neglecting the term $\left(\frac{\hbar\omega}{E_j - E_0}\right)^4 \approx 0$ in the denominator, one can rewrite the dispersion equation (12.19) in the form

$$\begin{aligned} 1 &= 2\kappa \sum_j \frac{|\langle\Phi_j|\hat{F}|\Phi_0\rangle|^2}{(E_j - E_0)^2}(E_j - E_0)\left(1 + \frac{(\hbar\omega)^2}{(E_j - E_0)^2}\right) = \\ &= 2\kappa \sum_j \frac{|\langle\Phi_j|\hat{F}|\Phi_0\rangle|^2}{(E_j - E_0)} + 2\kappa(\hbar\omega)^2 \sum_j \frac{\langle\Phi_j|\hat{F}|\Phi_0\rangle|^2}{(E_j - E_0)^3} \\ &\equiv 2\kappa\Sigma_1 + 2\kappa(\hbar\omega)^2\Sigma_3 . \end{aligned} \tag{12.39}$$

The last equality shows that the formula can express the vibration frequency

$$(\hbar\omega)^2 = \frac{(2\Sigma_1)^{-1} - \kappa}{\kappa\Sigma_3 / \Sigma_1} \equiv \frac{C}{B/\hbar^2}, \tag{12.40}$$

where the stiffness parameter and the mass parameter are

$$C = \frac{1}{2\Sigma_1} - \kappa, \tag{12.41}$$

$$B = \hbar^2\kappa\frac{\Sigma_3}{\Sigma_1}. \tag{12.42}$$

A phenomenological parameter is the strength κ of the residual force needed to solve the dispersion equation. When κ increases, the frequency of vibrations decreases to zero.

12.2.4 Including of pairing forces

The method described in the previous section gives results similar to experimental data only after including the pairing residual forces. Let us write the operator $\hat{F}$ in the second quantization formalism

$$\hat{F} = \sum_{\mu,\nu} \langle\mu|\hat{F}|\nu\rangle a_\mu^+ a_\nu \tag{12.43}$$

and then, using the Bogolubov-Valatin transformation, one expresses it by quasiparticle operators. Taking the two-quasiparticle states in their standard form

$$|\Phi_j\rangle \equiv |\lambda\mu\rangle = \alpha_\lambda^+ \alpha_\mu^+ |\Phi_0\rangle = \alpha_\lambda^+ \alpha_\mu^+ \prod_{\nu>0} (U_\nu + V_\nu a_\nu^+ a_{-\nu}^+)|0\rangle . \tag{12.44}$$

we can easily evaluate the matrix elements of $\hat{F}$:

$$\langle\Phi_j|\hat{F}|\Phi_0\rangle = \sum_{\mu\nu} \langle\mu|\hat{F}|\nu\rangle (U_\nu V_\nu + V_\mu U_\mu) . \tag{12.45}$$

The dispersion equation (12.19) takes now the following form:

$$\frac{1}{\kappa} = 2 \sum_{\mu\nu} \frac{|\langle\mu|\hat{F}|\nu\rangle|^2 (E_\mu + E_\nu)(U_\nu V_\mu + V_\nu U_\mu)^2}{(E_\mu + E_\nu)^2 - (\hbar\omega)^2} , \tag{12.46}$$

where $E_\nu = \sqrt{(e_\nu - \lambda)^2 + \Delta^2}$.

The mass parameters (12.36) will now be described by the formula:

$$B = 2\kappa^2 \hbar \sum_{\mu\nu} \frac{|\langle\mu|\hat{F}|\nu\rangle|^2 (E_\mu + E_\nu)}{[(E_\mu + E_\nu)^2 - (\hbar\omega)^2]^2} (U_\mu V_\nu + U_\nu V_\mu)^2 . \tag{12.47}$$

Depending on the type of vibration one is interested in, one substitutes for $\hat{F}$ various operators. For example, for shape oscillations, it is convenient to use multipole moment operators

$$\hat{F}_{\lambda\mu} = \hat{Q}_{\lambda\mu} . \tag{12.48}$$

For quadrupole oscillations with $\lambda = 2$, one has two types of operators $\hat{F}$: $\hat{F}_{20} = r^2 Y_{20}$ describes the vibrations which keep the axial symmetry of the nucleus and for $\mu = \pm 2$, those violating this symmetry $\hat{F}_{22} = r^2(Y_{22} + Y_{2-2})$. There are four types of octupole oscillations with $\lambda = 3$: for $\mu = 0$ oscillations corresponding to axially symmetric $\hat{F}_{30} = r^3 Y_{30}$ pear-like vibrations and for $\mu = 1, 2, 3$ three types of non-axial octupole excitations $\hat{F}_{31} = r^3(Y_{31} + Y_{3-1})$, $\hat{F}_{32} = r^3(Y_{32} + Y_{3-2})$, $\hat{F}_{33} = r^3(Y_{33} + Y_{3-3})$. Similar non-axial oscillation can appear at higher multipolarities, but in practice one considers the excitations with multipolarity $\lambda \leq 3$.

12.3 GIANT AND PYGMY RESONANCES

In the 40s of the last century, Baldwin and Klaiber first observed nuclei's giant resonances in high energy protons' scattering. The large cross-sections obtained with photons having about 15-20 MeV energy are understood as a collective nuclear response to the photon's electric dipole (E1) field. Such collective states have been predicted theoretically by Migdal as oscillations of the neutron versus proton distributions. All nuclei's giant dipole resonances (GDR) are above the particle emission threshold. In the early 60s, in the neutron capture experiments, a significant number of unbound states forming a bump around 5-8 MeV in the emitted γ-rays was also discovered. This bump, later called the pygmy dipole resonance (PDR), has energy approximately three times larger than the two quasiparticle excitations. A typical energy spectrum with both types of resonances is presented in Fig. 12.2.

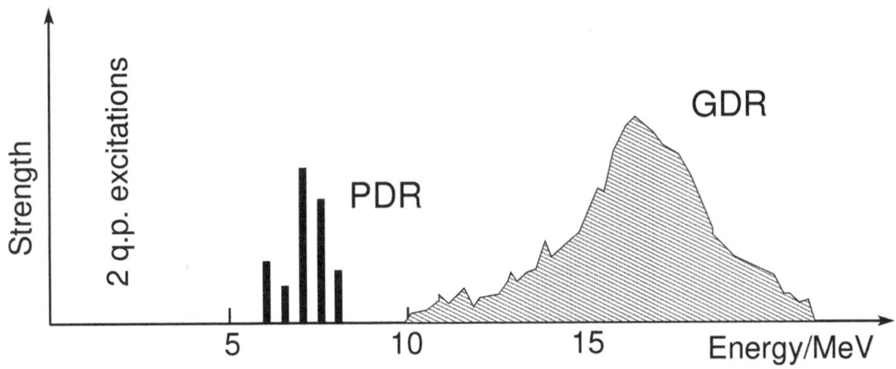

Figure 12.2 Schematic energy spectrum with the visible giant (GDR) and pygmy (PDR) dipole resonances.

Giant resonances are observed in nuclei from different mass regions, while the less pronounced pygmy resonances have been mainly found in neutron-rich nuclei. The description of the PDR as a collective excitation is based on a three-fluid model consisting of protons and neutrons in the same orbitals and the excess neutrons, which interact less strongly with the others.

In spherical nuclei, the photo-absorption cross-section corresponding to the giant dipole resonance (GDR) can be approximated by a Lorentz function

$$\sigma_{\mathrm{abs}}(E_\gamma) = \frac{\sigma_R E_\gamma^2 \Gamma_{\mathrm{GDR}}^2}{(E_\gamma^2 - E_{\mathrm{GDR}}^2)^2 - E_\gamma^2 \Gamma_{\mathrm{GDR}}^2}\,, \tag{12.49}$$

where E_{GDR} and Γ_{GDR} are the energy and width of the GDR, while σ_R is the maximal value of the cross-section. The excitation energy corresponding to the GDR in the medium and heavy nuclei is described by the following phenomenological formula:

$$E_{\mathrm{GDR}} = 31.2/A^{1/3} + 20.6/A^{1/6} \quad \mathrm{MeV}\,. \tag{12.50}$$

The energy pick corresponding to the GDR in the deformed nuclei is often split into two or three picks. The energies corresponding to the pick maxima are inversely proportional to the lengths of the nucleus in the direction of the principal axes. It gives a chance for an indirect measuring of the deformation of the nucleus.

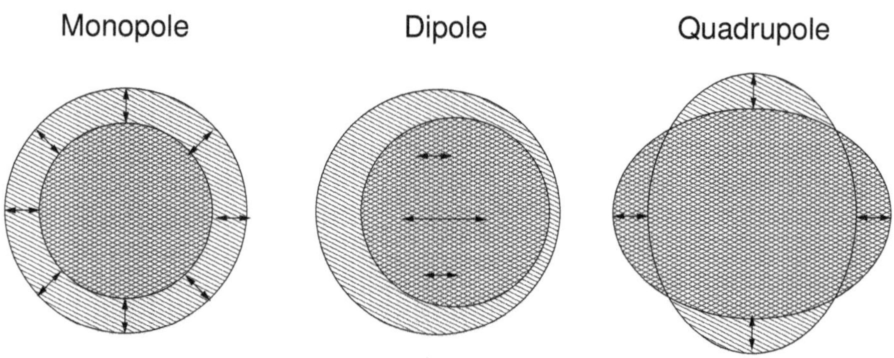

Figure 12.3 Three basic types of proton versus neutron oscillations

Apart from the giant dipole resonance, one also observes monopole and quadrupole resonances, which also correspond to oscillations of the proton versus neutron distributions like it is schematically shown in Fig. 12.3.

12.3.1 Goldhaber and Teller model

Let us assume for simplicity that the numbers of protons and neutrons are equal $Z = N = A/2$ and the displacement of each proton in the z-direction is $\zeta(t) = rP_1(\cos\theta)$, while that of neutron is in the opposite direction $(-\zeta(t) = -rP_1(\cos\theta)$. Both motions are taken in the opposite phase to keep the unchanged position of the nucleus mass center. Note that $rP_1(\cos\theta)$ is the dipole moment operator.

The kinetic energy related to this nucleon motion is

$$E_{\mathrm{kin}} = \sum_{\mathrm{protons}} \frac{1}{2} m_p \dot{\zeta}^2 + \sum_{\mathrm{neutrons}} \frac{1}{2} m_n \dot{\zeta}^2 \approx \frac{1}{2} A m \dot{\zeta}^2 = \frac{1}{2} M \dot{\zeta}^2\,, \tag{12.51}$$

where $m = (m_p + m_n)/2$ is the average nucleon mass and $M = Am$ is the mass of the whole nucleus.

Goldhaber and Teller have assumed that potential energy depends on the square of the difference of the proton and neutron densities $\rho_{sym} = \rho_p - \rho_n$

$$E_{sym} = \int \mathcal{V}_{sym}(\rho_{sym}^2) d^3r = \frac{1}{2} b_{sym} \int \frac{\rho_{sym}^2}{\rho_0} d^3r \,, \tag{12.52}$$

where ρ_0 is the nuclear matter density. For simplicity, one assumes in the following that $\rho_p(\vec{r}) = \rho_n(\vec{r}) = \frac{1}{2}\rho(\vec{r})$. The density can be expanded in the Taylor series around unperturbed positions of nucleons

$$\rho_{p(n)}(\vec{r} \mp \zeta \vec{u}_z) \approx \frac{1}{2}\left(\rho(\vec{r}) \mp \zeta \frac{\partial}{\partial z}\rho(\vec{r})\right) . \tag{12.53}$$

Inserting the above relation to Eq. (12.52), one obtains for the spherically symmetric density distribution

$$E_{sym} = \frac{b_{sym}}{\rho_0}\zeta^2 \int \left(\frac{\partial}{\partial z}\rho(\vec{r})\right)^2 d^3r = \frac{b_{sym}}{\rho_0}\zeta^2 \int \left(\cos(\theta)\frac{\partial}{\partial r}\rho(r)\right)^2 r^2 dr d\Omega \,. \tag{12.54}$$

After the integration over the 4π angle, the following formula for the symmetry energy is:

$$E_{sym} = \frac{4\pi}{6}\frac{b_{sym}}{\rho_0}\zeta^2 \int_0^\infty \left(\frac{\partial}{\partial r}\rho(r)\right)^2 r^2 dr = \frac{1}{2}K\zeta^2 \,, \tag{12.55}$$

where K is the constant, which is roughly proportional to the square of the nuclear radius (i.e., to $A^{2/3}$) as the density gradient is picked at the nuclear surface.

The sum of the kinetic (12.51) and the potential (12.55) symmetry energies give the classical Hamiltonian corresponding to the giant resonance:

$$\mathcal{H}_{class} = E_{kin} + E_{sym} = \frac{1}{2}M\dot{\zeta}^2 + \frac{1}{2}K\zeta^2 \,. \tag{12.56}$$

The frequency of the above classical oscillator

$$\omega_0 = \sqrt{\frac{K}{M}} \tag{12.57}$$

is proportional to $A^{-1/6}$ as one can deduce from Eqs. (12.51, 12.55).

After performing the quantization of the above harmonic oscillator Hamiltonian, like it was described in Sec. 6.1, one becomes its eigenenergies:

$$E_n = \frac{1}{2}(n+1)\hbar\omega_0 \,. \tag{12.58}$$

The difference between the first excited state energy and the ground state's energy $E_1 - E_0 = \hbar\omega_0$ corresponds to the giant dipole resonance energy. In real situations, protons' oscillation versus neutron density distributions are dumped, producing the final width of the GDR resonances.

Similar to the above model, one can also construct the other types of giant resonance assuming the corresponding type of vibrations taking $r^\lambda P_\lambda(\cos\theta)$ instead of the dipole $rP_1(\cos\theta)$ oscillations.

A better microscopic description of the giant resonances one can obtain, e.g., in the RPA type of calculations (see Sec. 12.2) assuming the difference of the proton and neutron multipole moment $Q_{\lambda 0} = r^\lambda P_\lambda(\cos\theta)$ operators as generators of the mean-field change. One can also extend this approach for the non-axial ($Q_{22} \sim r^2(Y_{22}(\theta,\varphi) + Y_{2-2}(\theta,\phi)$ or scissors $Q_{21} \sim r^2(Y_{21}(\theta,\varphi) + Y_{2-1}(\theta,\phi)$ modes.

The pygmy resonances, having energy around three times smaller than the giant ones, correspond to the vibration of the outer (valence) neutrons around the proton and neutron core. That is why they are observed chiefly in neutron-rich nuclei. Of course, the valence nucleons are not only neutrons but also protons. However, the amplitude of the proton oscillations cannot be so pronounced as that of neutrons because of the presence of the Coulomb barrier, which is absent for neutrons. This type of nuclear oscillation can be microscopically described by the cranking model assuming, e.g., the tilted in z direction Fermi surface model:

$$\hat{H}' = \hat{H} - \mu_p z_p + \mu_n z_n \,. \tag{12.59}$$

The relation between the Lagrange multipliers μ_p and μ_n has to be chosen to keep the nucleus mass centers unchanged. This approach is similar to those for the high-spin rotation in which the tilted Fermi surface was also introduced (Sec. 11.4, Fig. 11.6).

CHAPTER 13

Generalized Bohr Model

Models of rotation and vibrations of atomic nuclei, presented in Chapters 11 and 12 were based on a quantum mechanical approach to describe these phenomena. Because rotations and vibrations result from common nucleons' behavior, the theory of these types of multi-particle excitations is called a collective model . One can recall here a mathematically consistent derivation of the collective model describing simultaneous rotation and quadrupole vibrations, which is the basis of the Bohr model .

Based on the adiabatic approximation in the macroscopic approach or the microscopic calculations, the transition from a classic energy expression to the quantum Hamilton operator goes through given mathematical quantization rules to relevant operators attributing to classical quantities. Such rules, presented below, are not as strict as the generator coordinate method described in Chapter 15.

13.1 QUANTIZATION OF A CLASSICAL HAMILTONIAN

The method of quantizing a classic Hamiltonian should provide correspondence with classical mechanics. In addition, one assumes the following properties of the collective Hamilton operator, which should be:

- invariant due to time reversal,
- scalar,
- Hermitian,
- possess an appropriate symmetries depending on the system.

We will construct such a collective Hamiltonian being the sum of kinetic energy $\hat{T}$ and potential V

$$\hat{H}_{\text{coll}} = \hat{T} + V\,. \tag{13.1}$$

It is assumed that the kinetic energy operator should be quadratic in partial derivatives of collective q_i coordinates because classical kinetic energy is quadratic in collective velocities

$$T = \frac{1}{2}\sum_{ij} B_{ij}\dot{q}_i\dot{q}_j. \tag{13.2}$$

Using the quantization rules already proposed in the 20s by Pauli and Podolsky, one receives

$$\hat{T} = -\frac{\hbar^2}{2}\left(\sum_{ij}\frac{1}{D}\frac{\partial}{\partial q_i} D B_{ij}^{-1}\frac{\partial}{\partial q_j}\right) \tag{13.3}$$

The operator $\hat{T}$ is Hermitian with the following volume element

$$d\tau = D\,dq_1\cdots dq_n, \tag{13.4}$$

where

$$D = \sqrt{\det|B|}. \tag{13.5}$$

Solving the eigenproblem of the stationary collective Hamiltonian (13.1)

$$\hat{H}_{\text{coll}}\,\psi(q) = E\,\psi(q), \tag{13.6}$$

we will receive its eigenenergies and eigenstates.

By choosing the appropriate collective coordinates q_i, one can make the mass tensor diagonal and constant

$$B_{ij} = B_i\delta_{ij}. \tag{13.7}$$

The collective potential around the minimum can be approximated by a hyper-parabola

$$V(q) = \frac{1}{2}\sum_{ij} C_{ij}q_iq_j. \tag{13.8}$$

Proper scaling and rotation of q_i coordinates will cause that $C_{ij} = C_i\delta_{ij}$. One will then obtain the collective Hamiltonian in the form of a sum of the harmonic oscillators

$$\hat{H}_{\text{coll}} = -\frac{\hbar^2}{2}\sum_i\frac{1}{B_i}\frac{\partial^2}{\partial q_i^2} + \frac{1}{2}\sum_i C_i q_i^2 \tag{13.9}$$

having a characteristic spectrum of equal distances between eigenenergies equal to:

$$\hbar\omega_i = \hbar\sqrt{\frac{C_i}{B_i}}, \tag{13.10}$$

as already described in Chapter 12.

13.2 MICROSCOPIC COLLECTIVE BOHR HAMILTONIAN

As an example of a specific implementation of the collective model, let us consider the generalized Bohr model used to describe the rotation-oscillation spectrum. Only one of the microscopic methods of deriving this Hamiltonian, describing collective quadrupole excitations, i.e., of multipolarity $\lambda = 2$ will discussed here.

Suppose that single-particle motions in the nucleus are described by a Hamiltonian of an anisotropic harmonic oscillator with a correction term as in Nilsson potential (6.45)

$$\hat{H}_{\mathrm{sp}} = -\frac{\hbar^2}{2M}\hat{\Delta} + \frac{1}{2}M(\omega_x^2 x^2 + \omega_y^2 y^2 + \omega_z^2 z^2) + \hat{V}_{\mathrm{corr}}\,. \tag{13.11}$$

The surface of the constant potential, called the equipotential surface, has the shape of a spheroid described by Eq. (11.11). The half axes of the ellipsoid can be parameterized by introducing the quadrupole axial β and the non-axial γ nuclear deformation:

$$a = R_0(\gamma,\beta)\,[1 - k\cos(\gamma + 60^0)], \tag{13.12}$$

$$b = R_0(\gamma,\beta)\,[1 - k\cos(\gamma - 60^0)], \tag{13.13}$$

$$c = R_0(\gamma,\beta)\,[1 + k\cos\gamma]\,, \tag{13.14}$$

where

$$k = \sqrt{\frac{5}{4\pi}}\beta. \tag{13.15}$$

The parameters β and γ will be treated in the following as collective coordinates:

$$\begin{aligned} q_1 &\equiv \beta \in \langle 0, \sqrt{\tfrac{4\pi}{5}}\rangle\,, \\ q_2 &\equiv \gamma \in \langle 0.60^0\rangle\,. \end{aligned} \tag{13.16}$$

Let us transform now the Hamiltonian (13.11) going from the variables (x, y, z) to dimensionless stretched coordinates (6.55)

$$\begin{aligned} \xi &= x\sqrt{\frac{M\omega_x}{\hbar}}\,, \\ \eta &= y\sqrt{\frac{M\omega_y}{\hbar}}\,, \\ \zeta &= z\sqrt{\frac{M\omega_z}{\hbar}}\,. \end{aligned} \tag{13.17}$$

After this transformation, one gets a single-particle Hamiltonian in the form

$$\begin{aligned}
\hat{H}_{\rm sp} &= \frac{1}{2}\hbar\omega_0(-\hat{\Delta}_t+\rho^2)+ \qquad (13.18)\\
&+ \frac{1}{4}\hbar\omega_0\mathcal{B}\cos\mathcal{G}\left(-\frac{\partial^2}{\partial\xi^2}-\frac{\partial^2}{\partial\eta^2}+2\frac{\partial^2}{\partial\zeta^2}+\xi^2+\eta^2-2\zeta^2\right)+\\
&+ -\sqrt{\frac{3}{2}}\hbar\omega_0\mathcal{B}\sin\mathcal{G}\left(-\frac{\partial^2}{\partial\xi^2}+\frac{\partial^2}{\partial\eta^2}+\xi^2-\eta^2\right)+\\
&+ \hat{V}_{t\rm corr},
\end{aligned}$$

where

$$\mathcal{G} = \arctan\left(\frac{2\sin\gamma+\sqrt{\frac{5}{4\pi}}\beta\sin(2\gamma)}{2\cos\gamma-\sqrt{\frac{5}{4\pi}}\beta\cos(2\gamma)}\right) \qquad (13.19)$$

$$\mathcal{B} = \frac{2k\cos\gamma}{2\cos\mathcal{G}-\sqrt{\frac{5}{4\pi}}\beta\cos(\gamma+\mathcal{G})}. \qquad (13.20)$$

After diagonalization of the Hamiltonian $\hat{H}_{\rm sp}$ one receives a set of energies e_ν and one-particle functions $|\nu\rangle$ depending on the deformations β and γ

$$\hat{H}_{\rm sp}|\nu(\beta,\gamma)\rangle = e_\nu(\beta,\gamma)\,|\nu(\beta,\gamma)\rangle. \qquad (13.21)$$

We build a full, many-body Hamiltonian from the sum of single-particle and pairing Hamiltonians

$$\hat{H} = \sum_{i=1}^{A}\hat{H}_{\rm sp}(i)+\hat{H}_{\rm pair}. \qquad (13.22)$$

This Hamiltonian takes the form as in Eq. (9.1) in the second quantization representation. One approximately solves its eigenproblem using the BCS method, obtaining for even-even nuclei, the system of orthogonal states: ground state $|\Phi_0\rangle$ and two-quasiparticles excited states $\alpha^+_{\nu_1}\alpha^+_{\nu_2}|\Phi_0\rangle$ with energies respectively E_0 (9.33) and $E_{\nu_1\nu_2} = E_0+E_{\nu_1}+E_{\nu_2}$ (confer Eq. 9.48). Then, one solves the time-dependent Schroedinger equation

$$i\hbar\frac{\partial}{\partial t}|\psi\rangle = \hat{H}|\psi\rangle, \qquad (13.23)$$

assuming that $\hat{H}$ depends on the time t only via collective variables β, γ, and three Euler angles describing the orientation of the nucleus in the laboratory system. In the following, one will use the cranking model and the adiabatic approximation to obtain the solution of the above equation.

The energy of the nucleus, being the expected value of Hamiltonian (13.22) between the time-dependent functions ψ can be treated as the sum of collective kinetic and potential $V(\beta,\gamma)$ energy, obtained, e.g., by the Strutinsky method (Sec. 10.3)

$$\begin{aligned} E_{\rm coll} &= \frac{1}{2}(B_{\beta\beta}\dot{\beta}^2 + 2B_{\beta\gamma}\dot{\beta}\dot{\gamma} + B^2_{\gamma\gamma}\dot{\gamma}^2) + \\ &+ \frac{1}{2}(\mathcal{J}_x\omega_x^2 + \mathcal{J}_y\omega_y^2 + \mathcal{J}_z\omega_z^2) + V(\beta,\gamma) . \end{aligned} \tag{13.24}$$

The mass parameters $B_{q_iq_j}$ are a form analogous to (12.47) for $\omega = 0$

$$B_{q_iq_j} = 2\hbar^2 \sum_{\mu\nu} \frac{\langle\mu|\frac{\partial\hat{H}_{\rm sp}}{\partial q_i}|\nu\rangle\langle\nu|\frac{\partial\hat{H}_{\rm sp}}{\partial q_j}|\mu\rangle}{(E_\mu + E_\nu)^3}(U_\nu V_\mu + U_\mu V_\nu)^2 + \ldots , \tag{13.25}$$

where smaller terms containing time derivatives λ and Δ with respect to the coordinates q_i are omitted.

The moments of inertia $\mathcal{J}_{x_i}$, where $x_1 = x$, $x_2 = y$, $x_3 = z$ are given by the formula

$$\mathcal{J}_{x_i} = 2\hbar^2 \sum_{\mu\nu} \frac{|\langle\mu|\hat{j}_{x_i}|\nu\rangle|^2}{E_\mu + E_\nu}(U_\nu V_\mu - U_\mu V_\nu)^2 , \tag{13.26}$$

where $\hat{j}_{x_i}$ is x_i component of the single-particle angular momentum operator $\hat{\mathbf{j}} = \hat{\mathbf{l}} + \hat{\mathbf{s}}$.

Let us introduce rotational mass parameters

$$\begin{aligned} B_x &= \frac{\mathcal{J}_x}{4\beta^2 \sin^2\left(\gamma - \frac{2\pi}{3}\right)} , \\ B_y &= \frac{\mathcal{J}_y}{4\beta^2 \sin^2\left(\gamma + \frac{2\pi}{3}\right)} , \\ B_z &= \frac{\mathcal{J}_z}{4\beta^2 \sin^2\gamma} . \end{aligned} \tag{13.27}$$

It can be shown that for $\beta = 0$, $B_{\beta\beta} = B_{\gamma\gamma} = B_x = B_y = B_z$ and $B_{\beta\gamma} = 0$.

In the next step, one brings the classical collective Hamiltonian (13.1) to the corresponding quantum form using the Pauli-Podolsky quantization rules (see Eq. (13.3))

$$\hat{H}_{\rm coll} = \hat{T}_{\rm vib} + \hat{T}_{\rm rot} + V\,, \tag{13.28}$$

Kinetic vibrational energy operator has the form

$$\begin{aligned} \hat{T}_{\rm vib} &= -\frac{1}{2B_{\rm v}\sqrt{r}}\left\{\frac{1}{\beta^4}\frac{\partial}{\partial\beta}\left(\beta^4\sqrt{r}\frac{B_{\gamma\gamma}}{B_{\rm v}}\frac{\partial}{\partial\beta}\right) - \right. \\ &-\frac{1}{\beta^4}\frac{\partial}{\partial\beta}\left(\beta^3\sqrt{r}\frac{B_{\beta\gamma}}{B_{\rm v}}\frac{\partial}{\partial\gamma}\right) - \frac{1}{\beta\sin(3\gamma)}\frac{\partial}{\partial\gamma}\left(\sqrt{r}\sin(3\gamma)\frac{B_{\beta\gamma}}{B_{\rm v}}\frac{\partial}{\partial\beta}\right) + \\ &\left. + \frac{1}{\beta^2\sin(3\gamma)}\frac{\partial}{\partial\gamma}\left(\sqrt{r}\sin(3\gamma)\frac{B_{\beta\beta}}{B_{\rm v}}\frac{\partial}{\partial\gamma}\right)\right\}, \end{aligned} \tag{13.29}$$

where one uses the notation $r = B_x B_y B_z$ and $B_{\rm v} = \sqrt{B_{\beta\beta}B_{\gamma\gamma} - B_{\beta\gamma}^2}$.
The rotational energy operator has the form

$$\hat{T}_{\rm rot} = \frac{1}{2}\left\{\frac{\hat{I}_x^2}{\mathcal{J}_x} + \frac{\hat{I}_y^2}{\mathcal{J}_y} + \frac{\hat{I}_z^2}{\mathcal{J}_z}\right\}, \tag{13.30}$$

where operators $\hat{I}_{x_i} = \hat{I}_{x_i}(\varphi, \vartheta, \psi)$ are the collective angular momentum components. J_{x_i} are the moments of inertia.

The collective Hamiltonian $\hat{H}_{\rm coll}$ is the Hermitian operator, when one takes the volume element $d\tau$ in the form

$$d\tau = \beta^4\sqrt{r}\,B_{\rm v}|\sin(3\gamma)|\sin\vartheta\, d\beta\, d\gamma\, d\varphi\, d\vartheta\, d\psi\,, \tag{13.31}$$

where φ, ϑ, ψ are the Euler angles.

13.3 EXAMPLES OF SOLUTIONS

For some particular forms of the collective potential energy surface, one can find simple solutions of the Bohr Hamiltonian eigenproblem $\hat{H}_{\rm coll}$ (13.28). Two border cases will be presented below: vibrations around spherical shapes, and rotations and vibrations around deformed, axially symmetric shapes of nuclei.

The general Bohr's Hamiltonian eigenproblem solution requires more advanced numerical techniques and basic knowledge of group theory, so it will not be discussed in this manual.

13.3.1 Vibrations around the sphere

Suppose the nucleus is spherical, i.e., its ground state deformation is $\beta_{eq} = 0$. Then, the collective potential can be described roughly by a parabola

$$V(\beta,\gamma) = -\frac{1}{2}C\beta^2 . \qquad (13.32)$$

The Bohr Hamiltonian symmetry conditions for $\beta = 0$ assures that

$$B_{\beta\beta} = B_{\gamma\gamma} = B_x = B_y = B_z = \text{const} \equiv B, \quad \text{and} \quad B_{\beta\gamma} = 0 . \qquad (13.33)$$

The collective Hamiltonian then takes the form

$$\begin{aligned}\hat{H}_{\text{coll}} &= -\frac{1}{2B}\left[\frac{1}{\beta^4}\frac{\partial}{\partial\beta}\left(\beta^4\frac{\partial}{\partial\beta}\right) + \frac{1}{\beta^2\sin(3\gamma)}\frac{\partial}{\partial\gamma}\sin(3\gamma)\frac{\partial}{\partial\gamma} - \right. \\ &\quad -\frac{1}{4\beta^2}\left(\frac{\hat{I}_x^2}{\sin^2(\gamma-\frac{2\pi}{3})} + \frac{\hat{I}_y^2}{\sin^2(\gamma+\frac{2\pi}{3})} + \frac{\hat{I}_z^2}{\sin^2\gamma}\right) + \frac{1}{2}C\beta^2 .\end{aligned} \qquad (13.34)$$

It is a rather complicated form and not easy to discuss. However, it is good to remember that the Bohr Hamiltonian corresponds to the quadrupole moment $Q_{2\nu}$ vibration. Thus, the same Hamiltonian $\hat{H}_{\text{coll}}$ can be written in the laboratory system in the form of:

$$\hat{H}_{\text{coll}} = -\frac{1}{2B}\sum_{\mu}(-1)^{\mu}\hat{\Pi}_{\mu}\hat{\Pi}_{-\mu} + \frac{1}{2}C\sum_{\mu}(-1)^{\mu}q_{\mu}q_{-\mu} , \qquad (13.35)$$

where $\hat{\Pi}_{\mu} = -i\frac{\partial}{\partial q_{\mu}}$ is a momentum operator canonically coupled with the collective coordinate q_{μ}, being one of five components of the quadrupole tensor $Q_{2\nu}$. This form of $\hat{H}_{\text{coll}}$ can be easily brought to a five-dimensional harmonic oscillator:

$$\hat{H}_{\text{coll}} = \hbar\omega\sum_{\mu}\left(A_{\mu}^{+}A_{\mu} + \frac{1}{2}\right) , \qquad (13.36)$$

where

$$A_{\mu}^{+} = \frac{1}{\sqrt{2}}(\hat{\Pi}_{\mu} + iq_{\mu}) \quad \text{and} \quad A_{\mu} = (A_{\mu}^{+})^{+} . \qquad (13.37)$$

Eigenenergies of such five-dimensional harmonic oscillator $\hat{H}_{\text{coll}}$ are equidistant

$$E_N = \hbar\omega(N + \frac{5}{2}) \qquad (13.38)$$

and eigenfunctions are labeled with the number of quanta n_{μ} in one of five directions or by quantum numbers from the so-called physical base

$$\psi = |n_2\, n_1\, n_0\, n_{-1}\, n_{-2}\rangle = |N\,v\,a\,I\,M\rangle , \qquad (13.39)$$

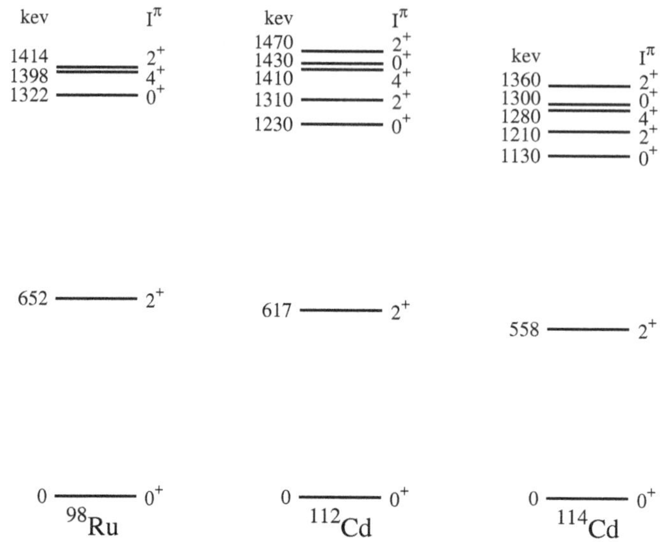

Figure 13.1 Spectrum of excited states of isotopes ^{92}Ru, ^{112}Cd, and ^{114}Cd.

where N is the total number of quanta, v – seniority number and a – an additional quantum number is unnecessary for the states with angular momentum $I \leq 6$. M is the third component of the angular momentum.

The energy spectrum of excited states similar to that of the spherical vibrators can be observed, e.g., in the ^{92}Ru, ^{112}Cd, and ^{114}Cd nuclei whose excited states spectra are shown in Fig. 13.1. Both spectra are characterized by almost perfectly equidistant energies, like those of a spherical oscillator.

13.3.2 Rotation-vibration model

Suppose the nucleus is quadrupole-deformed but has axial symmetry, i.e., $\beta_{\text{eq}} \neq 0$, $\gamma_{\text{eq}} = 0$. Then, the collective potential can be locally brought to the expression:

$$V(\beta,\gamma) = \frac{1}{8} C_{\beta\beta}\left(\beta^2 + \frac{\beta_{\text{eq}}^4}{\beta^2} - 2\beta_{\text{eq}}^2\right) + \frac{1}{2} C_{\gamma\gamma}\gamma^2, \tag{13.40}$$

to fulfill $V(\beta_{\text{eq}},0) = 0$ and $\frac{\partial^2}{\partial\beta^2} V(\beta_{\text{eq}},0) = C$.
In addition, let us assume that in the vicinity of the equilibrium point, the mass parameters are constant:

$$\begin{aligned} &B_{\beta\beta} = \text{const}, \quad && B_{\gamma\gamma} = \text{const}, \quad && B_{\beta\gamma} = 0, \\ &\mathcal{J}_x = \mathcal{J}_y = \text{const}, \quad && \mathcal{J}_z = 4\beta_{\text{eq}}^2 B_{\gamma\gamma}\gamma^2 . && \end{aligned} \tag{13.41}$$

This brings the Bohr's collective Hamiltonian (13.28-13.30) to a simpler separable form:

$$\begin{aligned}\hat{H}_{\rm coll} &= -\frac{1}{2B_{\beta\beta}}\frac{1}{\beta^4}\frac{\partial}{\partial\beta}\left(\beta^4\frac{\partial}{\partial\beta}\right)-\frac{1}{2B_{\gamma\gamma}\beta_{\rm eq}^2}\left[\frac{1}{\gamma}\frac{\partial}{\partial\gamma}\left(\gamma\frac{\partial}{\partial\gamma}\right)-\frac{I_z^2}{4\gamma^2}\right]+\\ &+\frac{1}{2\mathcal{J}}(I^2-I_z^2)+V(\beta,\gamma).\end{aligned} \tag{13.42}$$

The following formula gives its eigenenergies:

$$E_{n_\beta n_\gamma KI}=E_{n_\beta}^{(\beta)}+E_{n_\gamma}^{(\gamma)}+\frac{1}{2\mathcal{J}}[I(I+1)-K^2], \tag{13.43}$$

where

$$\begin{aligned}E_{n_\beta}^{\beta} &= \sqrt{\frac{C_{\beta\beta}}{B_{\beta\beta}}}\left(n_\beta+\tfrac{1}{2}\right),\\ E_{n_\gamma}^{\gamma} &= \sqrt{\frac{C_{\gamma\gamma}}{B_{\gamma\gamma}}}\left(n_\gamma+\tfrac{1}{2}\right),\end{aligned} \tag{13.44}$$

and K is a projection of angular momentum on the axis of symmetry of the nucleus. The Hamiltonian symmetry implies that the expression $n_\gamma - K/2$ must be a positive even number. The following quantum numbers mark the rotational-vibrational collective function

$$\psi=|n_\beta n_\gamma KIM\rangle, \tag{13.45}$$

where n_β, n_γ are quantum numbers for the β and γ vibrations, respectively, and M is the projection of the total angular momentum I on the z-axis of the laboratory system.

The energy spectrum of the excited states of the deformed nucleus will consist of a series of rotational states built on the vibrational states with different n_β, n_γ and K values listed below in columns:

n_β	0	1	0	2	1	0	0
n_γ	0	0	1	0	1	2	2
K	0	0	2	0	2	0	4

Such series of collective excited states are shown schematically in Fig. 13.2.

Analyzing the spectrum of excited states of ^{164}Er shown in Fig. 13.3, it is easy to identify the ground-state (g.s.) rotational band with increasing energies like $I(I+1)$. Similar bands are built at the first vibrational γ and β states. Two-particle excited states have energies of a few MeV and lie above the lowest vibrational collective states.

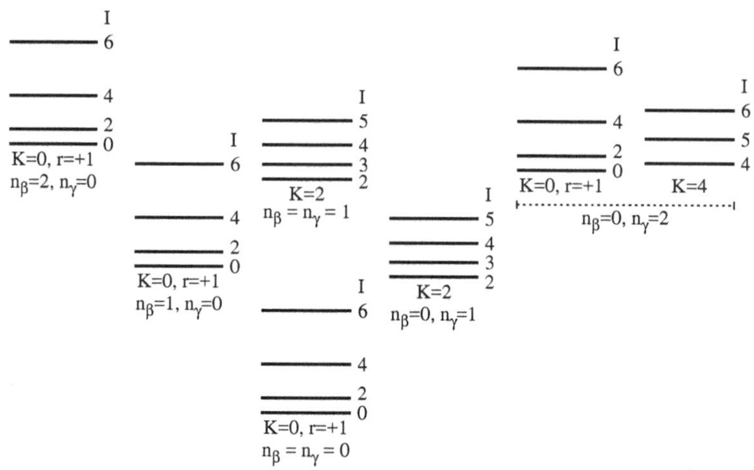

Figure 13.2 Schematic presentation of the vibration and rotation spectra.

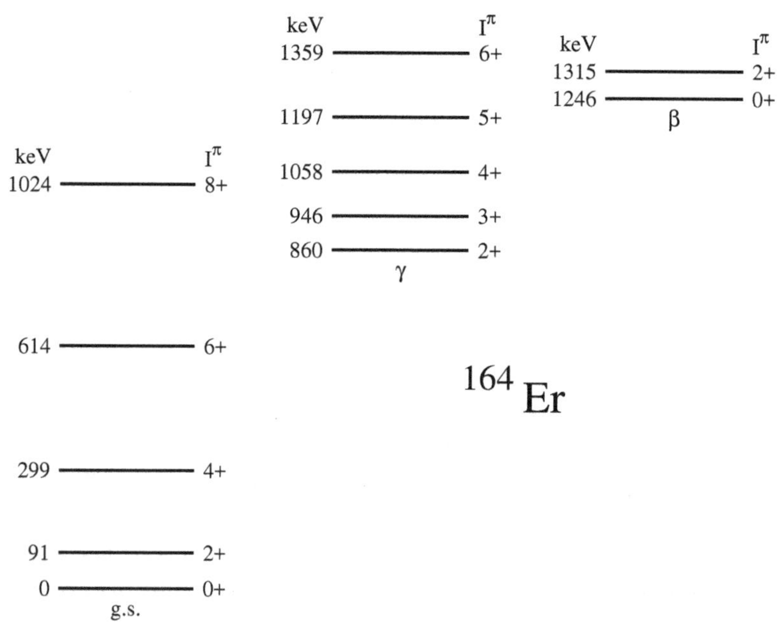

Figure 13.3 The lowest excited states of ^{164}Er.

Finding solutions of the general Bohr Hamiltonian is quite a complex problem that can only be solved by diagonalizing the Hamiltonian in an appropriate base or by numerical integration of the corresponding Schroedinger equation.

CHAPTER 14

Decays of atomic nuclei

Apart from around 3,300 known at present isotopes of nuclei, only 251 nuclides are stable. The other ones decay. The heaviest nuclei undergo a fission process or decay by alpha or, very rarely, by a cluster emission. Also, as mentioned in Sec. 5.1, the neutron-reach isotopes decay by the emission of an electron (β-decay) or by the capture of an electron from the lowest atomic orbit. The proton-reach isotopes may decay by a positron (β^+-decay) or proton emission. Of course, every decay process has to be energetically preferable: each physical system tries to reach the lowest possible energy. It is a general rule. Description of decays of excited nuclei or artificial nuclear reactions, e.g., heavy-ion synthesis of new elements, demands using more advanced statistical models. So, the book will not present the theory of such processes.

In this chapter, we will discuss spontaneous fission, alpha, proton, and cluster emission from the ground state of nuclei. The common aspect of these decays is a quantum-mechanical tunneling of the potential energy barrier. So, let us begin this chapter with a short reminder of such a tunneling process.

14.1 BARRIER PENETRATION IN THE WKB APPROXIMATION

Let us consider a particle of the mass m that strikes against a rectangular barrier of the width d and the height V_0. The particle energy (E) is smaller than the barrier height $E < V_0$, like it is shown in Fig. 14.1. The wave functions of incoming ($\psi_i(x)$; $x < 0$), out-coming ($\psi_o(x)$; $x > d$), and reflected ($\psi_r(x)$; $x < 0$) particle are described by a plane wave function $\sim \mathrm{e}^{ikx}$ and $\sim \mathrm{e}^{-ikx}$, respectively. Here, $k = p/\hbar = \sqrt{2E/m}/\hbar$ is a wave vector length. Inside the barrier ($0 \le x \le d$) the wave function $\psi(x)$ corresponds to

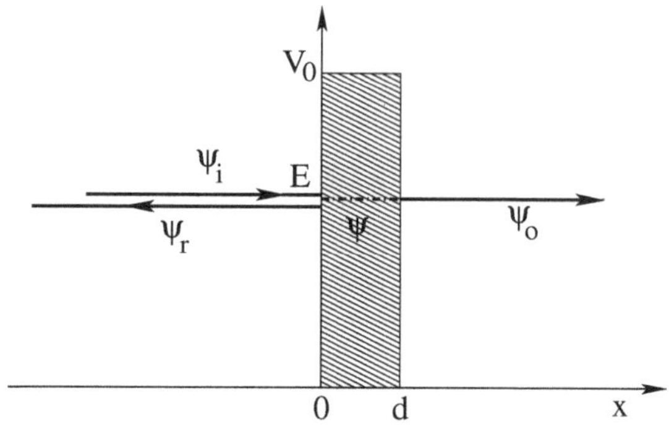

Figure 14.1 Schematic view of a particle striking a rectangular barrier.

eigenfunction of the following Schroedinger equation:

$$\left[-\frac{\hbar^2}{2m}\frac{d^2}{dx^2}+V_0\right]\psi(x)=E\,\psi(x) \tag{14.1}$$

and its eigenfunction is

$$\psi(x)=A\,\mathrm{e}^{-\sqrt{2(V_0-E)/m}/\hbar}\,, \quad \text{for} \quad 0\leq x\leq d\,. \tag{14.2}$$

The penetration probability $P(d)$ of such a single rectangular barrier is given by the following ratio:

$$P(d)=\frac{|\psi(d)|^2}{|\psi(0)|^2}=\exp\left[-\frac{2}{\hbar}\sqrt{2(V_0-E)/m}\,d\right] \tag{14.3}$$

In real situations, the potential energy barrier $V(x)$ is not rectangular, but one can always divide it into a series of N rectangular slabs of the height $V_i=V(x_i)$ and the width Δx like it is shown in Fig. 14.2. The probability (P) of penetration of such a more realistic one-dimensional barrier is given by the product of the penetration probability of each slab:

$$P=\prod_{i=1}^{N}P_i(d_i)=\exp\left[-\frac{2}{\hbar}\sum_{i=1}^{N}\sqrt{(2V_i-E)/m}\,\Delta x\right]. \tag{14.4}$$

The above expression goes to the Riemann integral when $\Delta x\to 0$, and the barrier penetration probability takes the form proposed in 1926 by Wentzl, Kramers, and Brillouin (WKB):

$$P=\exp\left[-\frac{2}{\hbar}\int_{V(x)\geq E}\sqrt{2(V(x)-E)/m}\,dx\right]. \tag{14.5}$$

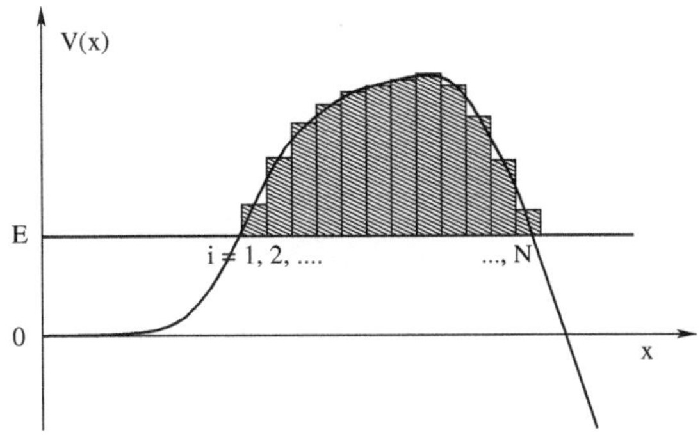

Figure 14.2 Penetration of a realistic barrier by a particle as a superposition of several (i=1,2,...., N) rectangular barrier transitions (see Eq. 14.4).

Here, the integration is performed in the x-range, where the potential energy ($V(i)$) is larger than the particle energy E. The formula (14.5) can be easily generalized for the case of a variable mass ($m(x)$), which happens when one deals with the inertial mass parameters (see Chapter 12).

In the following sections, the WKB approximation will be used to evaluate the spontaneous fission and alpha, cluster, or proton decay probabilities.

14.2 SPONTANEOUS FISSION

The neutron-induced nuclear fission of uranium nuclei was discovered in 1938 by Hahn and Strassmann and a month later was theoretically explained by Meitner and Frish as the partition of a nucleus into two parts. They have assumed that the nucleus becomes increasingly elongated after neutron injection and finally fissions like a biological cell. Until 1939, physicists believed that the atomic nucleus was spherical. The first spontaneous fission events of uranium nuclei were observed in 1940 by Flerov and Petrzhak, who have estimated its mean lifetime as $10^{16} - 10^{17}$ years. Both fission fragments are accelerated due to their Coulomb repulsions and reach a total kinetic energy of about 200 MeV, which was already predicted by Meitner and Frish in 1939. This sizeable kinetic energy and neutron emission has enabled the construction of nuclear power stations and, unfortunately, atomic bombs. Most trans-actinide nuclei split into two asymmetric fission fragments, while the heaviest one has a mass of around $A \approx 140$. Different neutron reach isotopes are produced, e.g., in the fission process induced by thermal neutrons (n_{th}):

$$^{235}\text{U} + 1\text{n}_{\text{th}} \longrightarrow {}^{140}\text{Ba} + {}^{94}\text{Kr} + 2\text{n} .$$

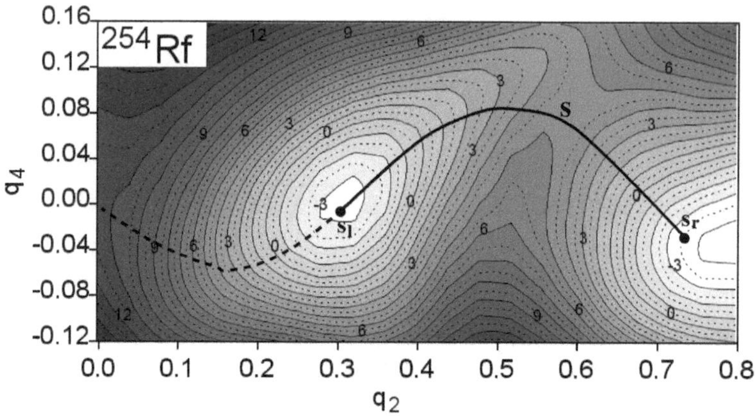

Figure 14.3 Two-dimensional cross-section of the four-dimensional PES of ^{254}Rf with marked static path to fission (solid blue line). Here, q_2 and q_4 are the deformation parameters responsible for the elongation and neck forming of the nucleus, respectively.

Some fission products are very radioactive and may have long decay lifetimes. It causes serious problems with reactor-waste management.

In this section, one concentrates on the spontaneous fission only, i.e., the fission of a nucleus from its ground state. This reaction may only happen due to the quantum mechanical tunneling of the so-called *fission barrier* . Let us construct a simple model. The fission barrier of a given nucleus can be found by analyzing the potential energy surface (PES) of the fissioning nucleus in the multidimensional $\{q_i\}$ deformation parameter space (6.82), like the one presented in Fig. 14.3, where the (q_2, q_4) cross-section of the four-dimensional PES of ^{254}Rf is shown. The classical, i.e., minimal energy, path to fission (thick line) begins at the ground-state equilibrium point (s_l) and runs up to the exit point (s_r), where the energy is equal to the ground-state energy E_0. The fission barrier obtained along such a path, where the energy has been minimized with respect to all deformation degrees of freedom except q_2, is presented in the l.h.s. panel of Fig. 14.4 as a function of the relative distance between the mass centers of both fragments $s = R_{12}/R_0$. The dashed line shows the phenomenological collective inertia B_{ss} (confer Sec. 12.2.2) that enters, through the action integral to the calculation of the spontaneous fission lifetime (see below). It is clear that along the fission path, each deformation parameter $q_i(s)$ is a function of the path length s. The classical energy

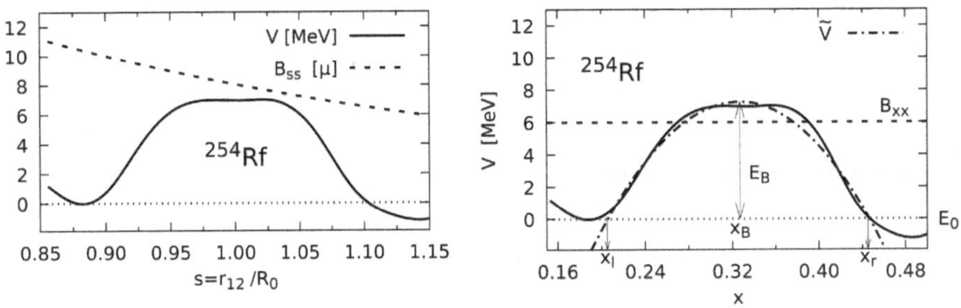

Figure 14.4 Potential energy barrier of ^{254}Rf along the path is shown in the l.h.s. panel as a function of the relative distance between the mass-centre of the fragments $s = R_{12}/R_0$. The dashed line shows the corresponding phenomenological mass parameter in the reduced mass units (μ). The potential as a function of the new coordinates x in which the mass parameter is constant ($B_{xx} = 6\mu$) is drawn in the r.h.s. panel. The dash-dotted line shows the approximation of this potential by two parabolas for $x < x_B$ and $x > x_B$.

$\mathcal{H}$ of the nucleus is the sum of the kinetic and potential V energies:

$$\mathcal{H} = \frac{1}{2}B_{ss}(s)\dot{s}^2 + V(s)\,, \tag{14.6}$$

where B_{ss} is the mass parameter and V the collective potential energy along the path s. A simple transformation from the s to a new x coordinate which conserves the kinetic energy:

$$x(s) = \int\limits_{s_{\rm sph}}^{s} \sqrt{\frac{B_{ss}(s')}{B}}\, ds'\,, \tag{14.7}$$

ensures that the mass parameter $B_{xx} = B$ corresponding to the new coordinate remains constant. In addition, one assumes here that $x = 0$ corresponds to the spherical nucleus shape ($R_{12}/R_0 = 0.75$). This transformation to a constant mass parameter is also valid in the quantum Hamiltonian, so one does not take into account the derivatives of the $B_{ss}(s)$ parameter. The potential $V[s(x)]$ in the new coordinate x, as shown in the r.h.s. panel of Fig. 14.4, can be approximated by two parabolas, one for each side of the barrier:

$$\widetilde{V}(x) = \begin{cases} V_{\rm sadd} - \frac{1}{2}C_l(x - x_{\rm B})^2 & \text{for } x < x_{\rm B}\,, \\ V_{\rm sadd} - \frac{1}{2}C_r(x - x_{\rm B})^2 & \text{for } x > x_{\rm B}\,, \end{cases} \tag{14.8}$$

how it is shown in the r.h.s. part of Fig. 14.4.

The spontaneous fission half-life is then given by:

$$T_{1/2}^{\rm sf} = \frac{\ln 2}{nP} \,. \tag{14.9}$$

Here P is the fission barrier tunneling probability (14.5)

$$P = \frac{1}{1+\exp\{2S(L)\}} \,, \tag{14.10}$$

and n is the number of strikes against the barrier in a time unit. The WKB action integral S along the fission path $L(x)$ is given by:

$$S(L) = \int_{s_l}^{s_r} \sqrt{\frac{2}{\hbar^2} B_{ss}[V(s) - E_0]}\, ds \approx \int_{-x_l}^{x_r} \sqrt{\frac{2B}{\hbar^2}[\widetilde{V}(x) - E_0]}\, dx \,. \tag{14.11}$$

The action integral for the fission barrier of height $E_{\rm B}$ composed by two inverted parabola like in Fig. 14.4 is equal to:

$$S = \frac{\pi}{2\hbar} E_{\rm B} \left(\sqrt{\frac{B}{C_l}} + \sqrt{\frac{B}{C_r}} \right) = \frac{\pi}{\hbar} E_{\rm B} \frac{\omega_l + \omega_r}{2\omega_l \omega_r} \equiv \frac{\pi}{\hbar} E_{\rm B}\, \tilde{\omega}^{-1} \,, \tag{14.12}$$

where $\omega_l = \sqrt{C_l/B}$ and $\omega_r = \sqrt{C_r/B}$ are the inverted harmonic oscillator frequencies.

For $S \gg 1$ the logarithm of the spontaneous fission half-lives (14.9) takes the form:

$$\log_{10}(T_{1/2}^{\rm sf}) = \frac{2\pi}{\hbar\tilde{\omega}} E_{\rm B} - \log_{10}[n \ln 2] \,. \tag{14.13}$$

Usually, one assumes that the number of assaults against the fission n equals the harmonic oscillator frequency corresponding to zero-point energy for the elongation (quadrupole) mode, approximately equal to 0.5 MeV.

The above theoretical estimate of $T_{1/2}^{\rm sf}$ is compared with the experimental data in Fig. 14.5 for nuclei from Th ($Z = 90$) to Ds ($Z = 110$). One has assumed here that the fission barrier height is equal to the difference between the LSD mass of nucleus (see Sec. 5.10) in the saddle point and the measured mass of the nucleus in its ground state $E_{\rm B} \approx M_{\rm sadd}^{\rm LSD} - M_{\rm g.s.}^{\rm exp}$. The quality of such approximation is visible in Fig. 14.6, where the experimental fission barrier heights of even-even actinide nuclei are compared with the above approximation. The average root-mean-square deviation of the measured and estimated barriers is here equal to 0.3 MeV only. It proves that the microscopic energy corrections at the saddle points are small. Additionally, in evaluating the spontaneous fission half-lives, one has assumed that the average inverted

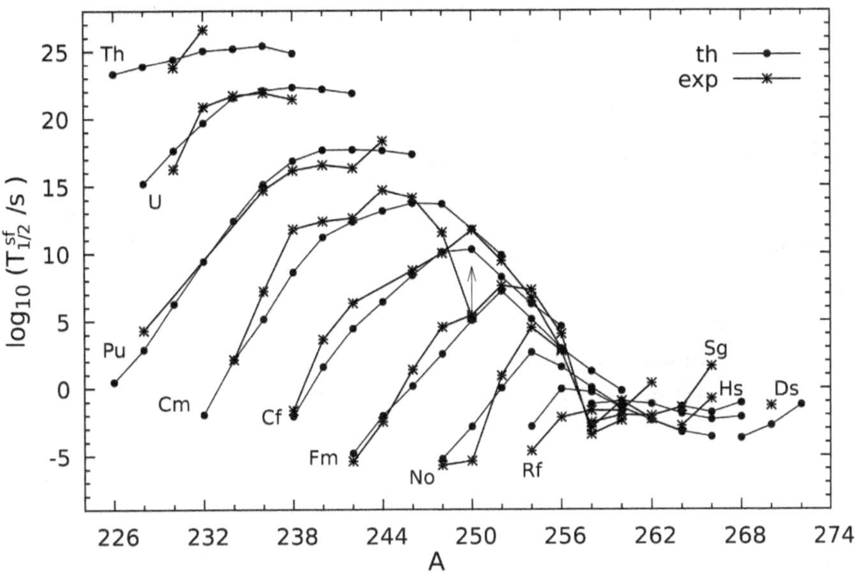

Figure 14.5 Logarithm of the spontaneous fission half-lives (points) predicted by formula (14.13) as a function of the mass number A. The experimental data (crosses) are taken from the NuDat 2.8 databases.

harmonic oscillator frequency $\hbar\tilde{\omega}$ is a quadratic function of the macroscopic fission barrier height:

$$\hbar\tilde{\omega} = a E_B^2 + b E_B + c \, .$$

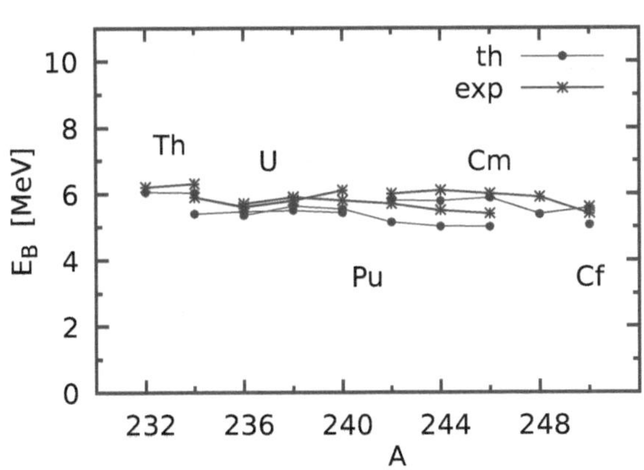

Figure 14.6 Experimental fission barrier heights compared with the barriers obtained in our simple approximation based on the Lublin-Strasbourg Drop (LSD) model.

The coefficients a, b, c are only the adjustable parameters when obtaining the estimates shown in Fig. 14.5. As one can see, such a simple one-dimensional WKB model reproduces the measured spontaneous fission lifetimes well.

14.3 ALPHA, CLUSTER, AND PROTON EMISSION

The lifetimes of nuclei decaying through a proton, α-particle or cluster (light nuclei) emission can be estimated quite accurately following the idea of Gamow, who in 1928 formulated the first alpha emission theory as the quantum-mechanical tunneling of the Coulomb barrier. Let us recall below the main assumption of the Gamow model .

Let us assume that a nucleus B decays into two parts: C, D:

$$^{A}B_{Z} \longrightarrow {}^{A_1}C_{Z_1} + {}^{A_2}D_{Z_2} ,$$

where the charges and the mass numbers of the emitted nucleus C and of the daughter nucleus D are denoted by (Z_1, A_1) and (Z_2, A_2), respectively. The decay by emission of a light-charged particle having energy E_k may be described as the quantum mechanical tunneling of the Coulomb barrier schematically presented in Fig. 14.7. One assumes here a rectangular nuclear

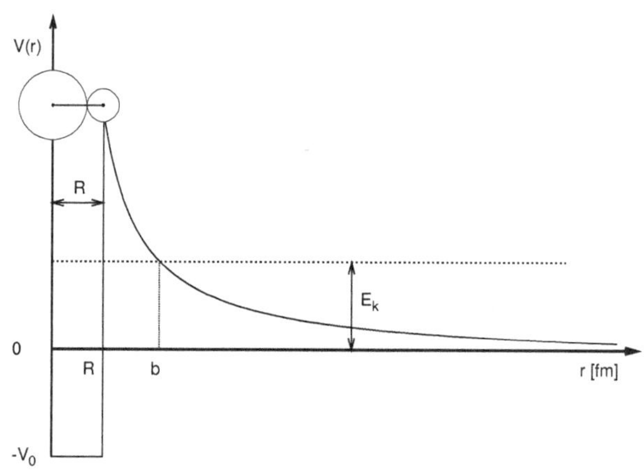

Figure 14.7 Potential barrier tunneling by an α particle.

potential of depth V_0 and radius R for the nuclear part of the potential energy. Outside the nucleus ($r > R$), only the Coulomb potential $V(r)$ is present. For simplicity, one assumes here that the parent nucleus and the emitted cluster are spherical. The exit point from the barrier is denoted by b. The energy of the emitted particle or light nucleus may be evaluated from the energy and momentum conservation rules:

$$\begin{cases} \frac{1}{2}M_C v_C^2 + \frac{1}{2}M_D v_D^2 = B(Z_1,A_1) + B(Z_2,A_2) - B(Z,A) \\ M_C v_C + M_D v_D = 0\,, \end{cases} \tag{14.14}$$

where $B(Z_i,A_i)$, M_i, and v_i are the binding energy (5.2), mass, and velocity of i-th nucleus, correspondingly. Assuming that the mass of nucleus M_i is proportional to its mass number A_i, one can easy evaluate the kinetic energy E_k of emitted particle ${}^{A_2}C_{Z_2}$:

$$E_k = \frac{1}{2}M_C v_C^2 = \frac{A_2}{A}[B(Z_1,A_1) + B(Z_2,A_2) - B(Z,A)]\,. \tag{14.15}$$

The decay half-life of a nucleus is then given by:

$$T_{1/2} = \frac{\ln 2}{\lambda}\,, \tag{14.16}$$

where λ is the decay width proportional to the Coulomb barrier tunneling probability P:

$$\lambda = nP\,. \tag{14.17}$$

Here, n is the number of assaults against the barrier. In the WKB approximation (14.5) this probability is:

$$P = \exp\left[-\frac{2}{\hbar}\int_R^b \sqrt{2\mu(V(r)-E_\mathrm{k})}dr\right], \tag{14.18}$$

where μ is the reduced mass of the emitted particle. The exit point from the barrier b corresponds to the place where the Coulomb potential is equal to the kinetic energy E_k:

$$b = \frac{e^2 Z_1 Z_2}{E_\mathrm{k}} \tag{14.19}$$

with e^2=1.44 fm · MeV being the square of the elementary charge.

The probability of tunneling of the Coulomb barrier of a spherical nucleus is given by:

$$P = \exp\left\{-\frac{2}{\hbar}\sqrt{2\mu Z_1 Z_2 e^2 b}\cdot\left[\arccos\sqrt{R/b} - \sqrt{R/b-(R/b)^2}\right]\right\}. \tag{14.20}$$

Here, $R = r_0(A_1^{1/3} + A_2^{1/3})$ is the radius of the square well shown in Fig. 14.7. In addition, one assumes that the emitted particle is, in its ground state, in a square well with relatively high walls like presented in Fig. 14.7. This assumption allows to use of the number of assaults per time-unit (14.17) against the barrier corresponding to the ground state frequency of an infinite square well:

$$n \approx \frac{\pi \hbar}{2\mu R^2} \,. \tag{14.21}$$

Please note that this model's only adjustable parameter is the radius constant r_0 of the square well radius. A least-square fit to all known α-decay lifetimes of even-even nuclei gives $r_0 = 1.221$ fm. In Fig. 14.8, the logarithm of the half-lives for the α-particle emission $\log_{10}(T^{\alpha}_{1/2})$ for even-even heavy and super-heavy nuclei are compared with the experimental data taken from the NuDat 2.8 database. It is shown that the estimates obtained within the described above simple model reproduce the data, if available, quite well.

Similarly, good accuracy for the reproduction of the decay half-lives for cluster and proton decays , presented in Figs. 14.9 and 14.10, are obtained without any adjustment of the radius constant r_0. All experimental half-lives

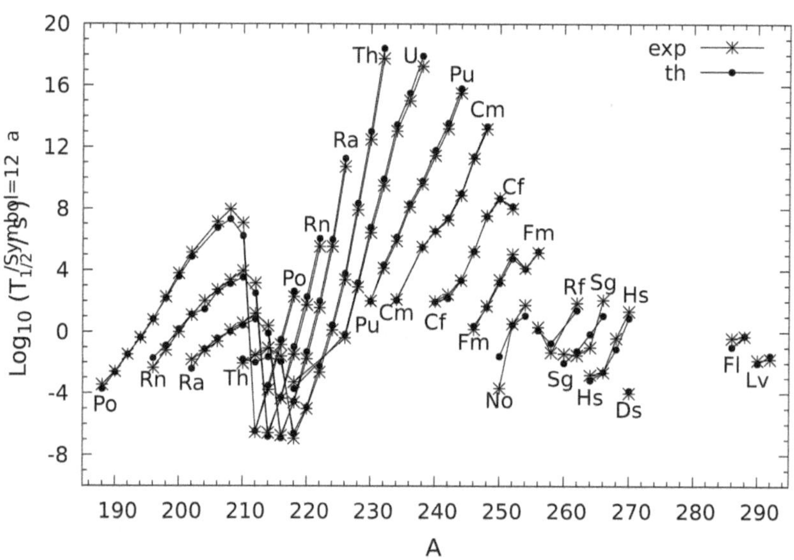

Figure 14.8 Theoretical estimates of α decay half-life time of even-even heavy nuclei compared with the data (crosses) as a function of the mass number of the emitter. The experimental values of $T^{\alpha}_{1/2}$ and Q_{α} decay energies are taken from the NuDat 2.8 databases.

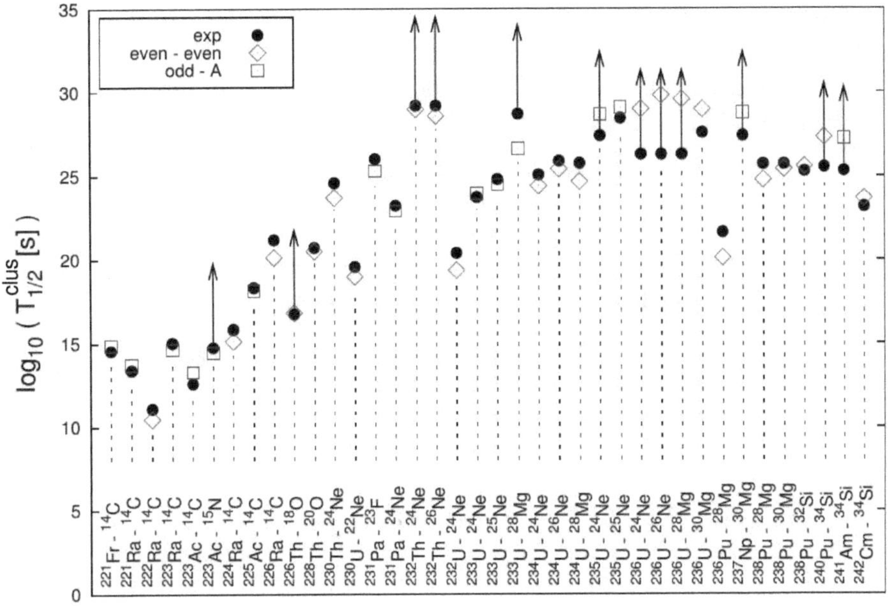

Figure 14.9 Logarithm of cluster emission half-lives. The arrows mean that the lower limit is measured only.

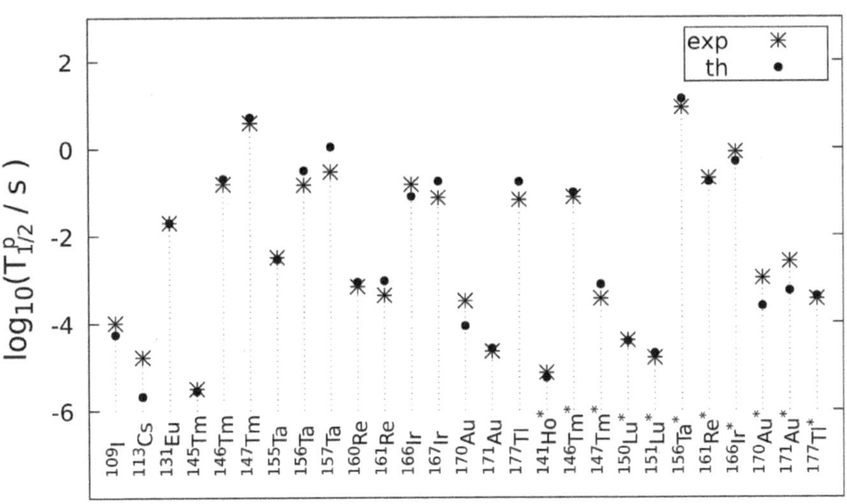

Figure 14.10 Logarithm of proton emission half-lives

and their estimates presented in Figs. 14.8 – 14.10 are in seconds. Note that one year corresponds to $10^{7.5}$ seconds.

To describe the decay of nuclei having an odd mass number A, one introduces an additional hindrance factor h:

$$T^{\alpha}_{1/2} = \frac{\ln 2}{\lambda} \cdot 10^{h} . \tag{14.22}$$

A fit to the alpha-decay lifetimes of odd-even and even-odd nuclei gives $h = 0.216$. For odd-odd nuclei, the hindrance factor is doubled.

CHAPTER 15

Generator coordinate method and nuclear dynamics

Two types of coordinates describe nuclear phenomena:

- single-particle coordinates:
$$\mathbf{x} = \{x_1, x_2, \ldots, x_A\}, \tag{15.1}$$
where x_i means the coordinates of the location $\mathbf{r}_i$, spin $\hat{\sigma}_i$ and isospin $\hat{\tau}_i$ for ith nucleon, giving total 6A variables for one type of particle;
- collective coordinates for the entire nucleus $\{q\}$, chosen depending on the phenomenon one would like to describe.

The nature and number of collective coordinates change in different models. These can be, e.g., deformation parameters $\varepsilon_\lambda = \{\varepsilon, \varepsilon_3, \varepsilon_4, \ldots\}$. In most theoretical approaches, collective coordinates are usually independent of single-particle coordinates. Such a procedure increases the number degrees of freedom needed to describe the whole nucleus. It causes an artificial magnification of the number of coordinates, which can lead to non-physical effects predicted by such models.

The problem of constructing collective coordinates from single-particle variables has troubled nuclear theorists for a long time. Various conditions were imposed on the coordinates of nucleons or the function describing the entire nucleus, whose eigenstate, e.g., in the Born-Oppenheimer approximation, is assumed as a product of the function χ dependent on the single-particle coordinates and the collective one Φ:

$$\psi(\mathbf{x}, q) = \chi(\mathbf{x}; q)\Phi(q). \tag{15.2}$$

The total energy operator of the nucleus consists then of the collective part $\hat{H}_{\rm coll}$ and internal one $\hat{H}$, described, e.g., by the sum of single-particle Hamiltonians dependent directly on nucleon degrees of freedom and parametrically on the collective coordinates.

The eigenequation of the internal Hamiltonian

$$\hat{H}(\mathbf{x};q)\chi(\mathbf{x};q) = E(\mathbf{x};q)\chi(\mathbf{x};q) \tag{15.3}$$

can be solved, e.g., by the BCS or Hartree-Fock method

$$\chi(\mathbf{x};q) = |BCS\,(\mathbf{x};\varepsilon_\lambda)\rangle \quad \text{and} \quad E = E_{\rm BCS}\,. \tag{15.4}$$

By diagonalizing the collective Hamiltonian $\hat{H}_{\rm coll}$, corresponding to the classical kinetic and potential energies of a given system, one gets the collective eigenfunctions of Φ_α and eigenenergies $\mathcal{E}_\alpha$

$$\hat{H}_{\rm coll}(q)\,\Phi_\alpha(q) = \mathcal{E}_\alpha\,\Phi_\alpha(q)\,. \tag{15.5}$$

To obtain the mean value of a certain observable Q, one has to evaluate the average value of the corresponding operator $\hat{Q}$ between functions (15.2) by integrating over all single-particle $\{\mathbf{x}\}$ and collective $\{q\}$ coordinates

$$Q = \int\!\!\int \psi^\star \hat{Q}\psi\, d\mathbf{x}dq = \int \Phi^\star(q)\left[\int \chi^\star(\mathbf{x};q)\,\hat{Q}\chi(\mathbf{x};q)\,d\mathbf{x}\right]\Phi(q)\,dq\,. \tag{15.6}$$

One can do it in two steps: one evaluates first the integral over the single-particle variables and stores in many grid points q:

$$Q^{\rm st}(q) = \int \chi^\star(\mathbf{x};q)\,\hat{Q}\chi(\mathbf{x};q)\,d\mathbf{x} \equiv \langle q|\hat{Q}|q\rangle\,, \tag{15.7}$$

and then one integrates it numerically with respect to collective coordinates

$$Q = \int \Phi^*(q)\,Q^{\rm st}(q)\,\Phi(q)\,dq\,. \tag{15.8}$$

Such a way of proceeding reflects better the dynamical nature of nuclear observables than just taking their static values $Q^{\rm st}(q_0)$ in the ground-state q_0, i.e., in the minimum energy concerning q. The Born-Oppenheimer method omits the problem of selecting the collective and related to them single-particle coordinates.

Hill and Wheeler have found a solution to this problem. They have proposed the Generator Coordinate Method (GCM), in which the function ψ

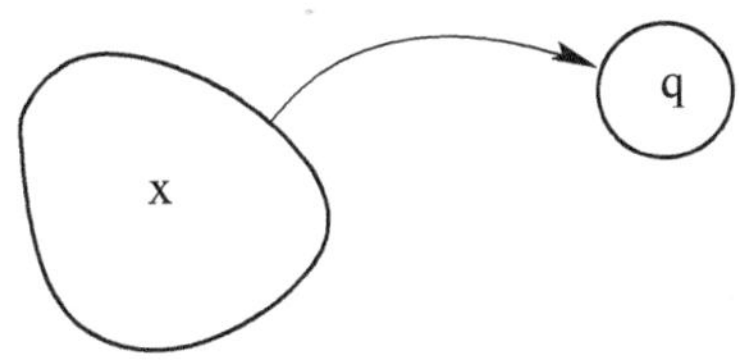

Figure 15.1 Projecting of single-particle coordinates onto the collective ones.

describing the entire nucleus is postulated in the form of a convolution of χ functions, hereinafter referred to as generating functions

$$\psi(\mathbf{x}) = \int_{-\infty}^{+\infty} f(q)\chi(\mathbf{x};q)\,dq\,, \tag{15.9}$$

where $f(q)$ is a weight function. The form of these weight functions can be obtained by minimizing the mean value of Hamiltonian

$$\delta\frac{\langle\psi|\hat{H}|\psi\rangle}{\langle\psi|\psi\rangle} = 0 \tag{15.10}$$

with respect to the different choices of these functions. This procedure leads to the Hill-Wheeler set of equations, which is hard to solve in practical applications. However, there are methods of an approximate solution of this variational problem. One of them discussed below, consists of the Gaussian approximation of the overlap function.

Let us first evaluate the average value of a certain Hermitian operator $\hat{Q}$ (e.g. $\hat{H}$) in the state ψ (15.9)

$$Q = \langle\psi|\hat{Q}|\psi\rangle\,. \tag{15.11}$$

Using the Gauss approximation when evaluating this average value, one can obtain formulas for the corresponding collective operator $\hat{Q}_{\rm coll}$ (e.g., $\hat{H}_{\rm coll}$). Such collective operators will consist of the kinetic and potential parts. One can separate the integral (15.11) into strongly and weakly dependent on variables q and q' parts:

$$\begin{aligned}\langle\psi|\hat{Q}|\psi\rangle &= \int_{-\infty}^{+\infty} dq \int_{-\infty}^{+\infty} f^\star(q)\left[\int \chi^\star(\mathbf{x};q)\,\hat{Q}\chi(\mathbf{x};q')\,d^A\mathbf{x}\right] f(q')\,dq' = \\ &= \int_{-\infty}^{+\infty} dq \int_{-\infty}^{+\infty} f^\star(q)\int \chi^\star(\mathbf{x};q)\chi(\mathbf{x};q')\,d^A\mathbf{x}\,\frac{\int \chi^\star(\mathbf{x};q)\,\hat{Q}\chi(\mathbf{x};q')\,d^A\mathbf{x}}{\int \chi^\star(\mathbf{x};q)\chi(\mathbf{x};q')\,d^A\mathbf{x}}\,f(q')\,dq'\,.\end{aligned} \tag{15.12}$$

The integration with respect to the coordinates q and q' runs all over the range of their variability, which one symbolically denotes here by $\pm\infty$.
The integral

$$\int \chi^\star(\mathbf{x};q)\chi(\mathbf{x};q')d^A\mathbf{x} \equiv \langle q|q'\rangle \tag{15.13}$$

which holds the main dependence on q coordinates, is called the overlap of the functions χ. The other term in Eq. (15.12), being the quotient of two integrals, depends rather weakly on q and is called the reduced overlap.

In the Gaussian Overlap Approximation (GOA) one assumes that the integral (15.13) can be approximate by a Gaussian function. In one dimension $\{q\} = q$ it can be written as

$$\langle q|q'\rangle = \mathrm{e}^{-\frac{1}{2}\gamma(\bar{q})(q-q')^2}\,. \tag{15.14}$$

After entering the following scaled coordinates:

$$\alpha = \int\limits_{q_0}^{q} \sqrt{\frac{\gamma(q')}{\gamma_0}}\,dq' \tag{15.15}$$

the overlap function takes the form

$$\mathcal{N}(\alpha,\alpha') = \langle\alpha|\alpha'\rangle = \mathrm{e}^{-\frac{1}{2}\gamma_0(\alpha-\alpha')^2}\,, \tag{15.16}$$

where γ_0 is a constant.

Then one can introduce the function $\mathcal{N}^{(\frac{1}{2})}(\alpha,\xi)$ such that

$$\mathcal{N}(\alpha,\alpha') = \int\limits_{-\infty}^{+\infty} \mathcal{N}^{(\frac{1}{2})}(\alpha,\xi)\mathcal{N}^{(\frac{1}{2})}(\alpha',\xi)\,d\xi\,, \tag{15.17}$$

and has the following form

$$\mathcal{N}^{(\frac{1}{2})}(\alpha,\xi) = \left(\frac{2}{\pi}\right)^{\frac{1}{4}} \exp\left\{-(\alpha-\xi)^2\right\}. \tag{15.18}$$

Hence $\mathcal{N}(\alpha,\alpha')$ is the Gauss function and $\gamma_0 = 1$.
After inserting the above expressions into the formula (15.12) for the average values of the $\hat{Q}$ operator, one has

$$\langle\psi|\hat{Q}|\psi\rangle = \int\limits_{-\infty}^{+\infty}\int\limits_{-\infty}^{+\infty} g^\star(\alpha)\,Q(\alpha,\alpha')\,g(\alpha')\,\mathcal{N}(\alpha,\alpha')\,d\alpha\,d\alpha'\,, \tag{15.19}$$

where $g(\alpha)$ comes from the transformation of the weight function f:

$$f(q(\alpha)) \equiv g(\alpha)\,. \tag{15.20}$$

The reduced overlap function for the operator $\hat{Q}$ appearing in Eq.(15.12)

$$\frac{\langle q|\hat{Q}|q'\rangle}{\langle q|q'\rangle} \equiv Q(q(\alpha), q'(\alpha')) = Q(\alpha, \alpha'), \tag{15.21}$$

which shows a rather weak dependence on α, α' can be developed into a Taylor series around a certain intermediate point $\xi = (\alpha + \alpha')/2$. In addition, let us assume that one is dealing with more than one α (or q) parameter and $\alpha \to \{\alpha_\mu\}$, where $\mu = 1, 2, ..., n$. Then

$$\begin{aligned} Q(\alpha, \alpha') &= Q(\xi, \xi) + \left.\frac{\partial Q}{\partial \alpha_\mu}\right|_\xi (\alpha_\mu - \xi_\mu) + \left.\frac{\partial Q}{\partial \alpha'_\mu}\right|_\xi (\alpha'_\mu - \xi_\mu) \\ &+ \frac{1}{2}\left\{ \left.\frac{\partial^2 Q}{\partial \alpha_\mu \partial \alpha_\nu}\right|_\xi (\alpha_\mu - \xi_\mu)(\alpha_\nu - \xi_\nu) + 2\left.\frac{\partial^2 Q}{\partial \alpha_\mu \partial \alpha'_\nu}\right|_\xi (\alpha_\mu - \xi_\mu)(\alpha'_\nu - \xi_\nu) \right. \\ &+ \left. \left.\frac{\partial^2 Q}{\partial \alpha'_\mu \partial \alpha'_\nu}\right|_\xi (\alpha'_\mu - \xi_\mu)(\alpha'_\nu - \xi_\nu) \right\} + \ldots . \end{aligned} \tag{15.22}$$

Here, one has adopted the convention of summing over repeating μ and ν indexes. Differentiating $Q(\alpha, \alpha')$ (15.21) over α_μ, α_ν, α'_μ, and α'_ν one can evaluate the corresponding expansion coefficients. For assumed form (15.18) of the function $\mathcal{N}^{\frac{1}{2}}(\alpha, \xi)$ it can be proven that

$$\begin{aligned} \mathcal{N}^{(\frac{1}{2})}(\alpha, \xi)(\alpha_\nu - \xi_\nu) &= \frac{1}{2}\frac{\partial}{\partial \xi_\nu}\mathcal{N}^{(\frac{1}{2})}(\alpha, \xi), \\ \mathcal{N}^{(\frac{1}{2})}(\alpha, \xi)(\alpha_\nu - \xi_\nu)(\alpha_\mu - \xi_\mu) &= \frac{1}{4}\frac{\partial^2}{\partial \xi_\mu \partial \xi_\nu}\mathcal{N}^{(\frac{1}{2})} + \frac{1}{2}\delta_{\mu\nu}. \end{aligned} \tag{15.23}$$

Let us denote

$$\Phi(\xi) \equiv \int_{-\infty}^{+\infty} \mathcal{N}^{(\frac{1}{2})}(\xi, \alpha') g(\alpha') d\alpha' \tag{15.24}$$

and use the symmetries of the development coefficients:

$$\left.\frac{\partial^2 Q(\alpha,\alpha')}{\partial\alpha'_\nu\partial\alpha'_\mu}\right|_\xi \equiv Q_{\nu\mu}, \qquad \left.\frac{\partial Q(\alpha,\alpha')}{\partial\alpha'_\mu}\right|_\xi \equiv Q_\mu,$$
$$\left.\frac{\partial^2 Q(\alpha,\alpha')}{\partial\alpha_\nu\partial\alpha'_\mu}\right|_\xi \equiv {}_\nu Q_\mu, \qquad \left.\frac{\partial Q(\alpha,\alpha')}{\partial\alpha_\mu}\right|_\xi \equiv {}_\mu Q, \tag{15.25}$$
$$\left.\frac{\partial^2 Q(\alpha,\alpha')}{\partial\alpha_\nu\partial\alpha_\mu}\right|_\xi \equiv {}_{\nu\mu} Q, \qquad Q(\xi,\xi) \equiv Q_0 .$$

After inserting the relations (15.22 - 15.25) to the right side of Eq. (15.19) and after returning to the variables q one gets the following result

$$\langle\psi|\hat{Q}|\psi\rangle = \int_{-\infty}^{+\infty} \Phi^\star(q)\hat{Q}_{\rm coll}\Phi(q)\sqrt{\det\gamma}\, d^n q, \tag{15.26}$$

where the operator $\hat{Q}_{\rm coll}$ consists of kinetic (with derivatives over q) and potential parts

$$\hat{Q}^{\rm coll} = \hat{Q}^T + \hat{Q}^V \tag{15.27}$$

and it is the projection operator of the $\hat{Q}$ operator, which acts in the many-body fermionic space onto the collective (i.e., generating) coordinate space. The potential part of the $\hat{Q}_{\rm coll}$ operator

$$\hat{Q}^V = \langle q|\hat{Q}|q\rangle - \frac{1}{2}\gamma_{ij}^{-1}\, {}_jQ_i \tag{15.28}$$

contains the average value of the $\hat{Q}$ operator

$$Q_0 = \langle q|\hat{Q}|q\rangle, \tag{15.29}$$

and an additional term that takes into account the collective correlations already present in the state $|q\rangle$

$$\mathcal{E}_0 = \frac{1}{2}\gamma_{ij}^{-1}\, {}_jQ_i . \tag{15.30}$$

This term is frequently called the zero-point vibrational correction. The kinetic part of the $\hat{Q}_{\rm coll}$ operator is given by

$$\hat{Q}^T = \frac{1}{2\sqrt{\det\gamma}}\frac{\partial}{\partial q_m}\sqrt{\det\gamma}\,\mathbf{M}^{-1}_{ml}\frac{\partial}{\partial q_l}, \tag{15.31}$$

where $\mathbf{M}_{ml}$ plays the role of a mass tensor and is equal to:

$$\mathbf{M}^{-1}_{ml} = \frac{1}{2}\gamma^{-1}_{mk}\left(2\,{}_kQ_j - \frac{\partial}{\partial q_k}Q_j + \Gamma^{i}_{kj}\,Q_i\right)\gamma^{-1}_{jl}\,. \tag{15.32}$$

Γ^i_{kj} stays for the Christoffel symbols and the matrix elements ${}_jQ_k$ are equal to:

$${}_jQ_k = \langle q|\frac{\overleftarrow{\partial}}{\partial q_j}(\hat{Q} - Q_0)\frac{\overrightarrow{\partial}}{\partial q_k}|q\rangle_{\mathrm{L}}\,. \tag{15.33}$$

It is a matrix element without a diagonal part (linked). Arrows point here the direction of differentiation operators.
Similarly, the width of the overlap function is equal

$${}_i\gamma_j = \langle q|\frac{\overleftarrow{\partial}}{\partial q_i}\frac{\overrightarrow{\partial}}{\partial q_j}|q\rangle \tag{15.34}$$

and

$$Q_j = \langle q|\hat{Q}\frac{\overrightarrow{\partial}}{\partial Q_j}|q\rangle\,. \tag{15.35}$$

The form of the collective operator (15.27) enables the calculation of dynamical values of observables represented by Hermitian operators $\hat{Q}$. In the case when $\hat{Q}$ is a Hamiltonian operator of independent quasiparticles, e.g., $\hat{Q} \to \hat{H} = \hat{H}_0 + \hat{H}_{\mathrm{pair}}$, it is convenient to take the BCS wave function as a generating function χ. Then, one receives the collective Hamiltonian in a form similar to the operator by the Pauli-Podolsky quantization method (13.1 - 13.5) of classical energy

$$\hat{H}_{\mathrm{coll}} = -\frac{1}{2\sqrt{\det\gamma}}\frac{\partial}{\partial q_i}\sqrt{\det\gamma}\,B^{-1}_{ij}\frac{\partial}{\partial q_j} + V_{\mathrm{coll}}\,. \tag{15.36}$$

The mass parameters B_{ij} in the above equation have the following form:

$$B^{-1}_{ij} = \frac{1}{2}\gamma^{-1}_{ik}({}_kh_l)\gamma^{-1}_{lj}\,, \tag{15.37}$$

where

$${}_kh_l = \langle q\left|\frac{\overleftarrow{\partial}}{\partial q_k}(\hat{H} - E_{\mathrm{BCS}})\frac{\overrightarrow{\partial}}{\partial q_l}\right|q\rangle\,. \tag{15.38}$$

Note, that the $\mathcal{E}_0$ (15.30) correction

$$\mathcal{E}_0 = \frac{1}{2}\gamma^{-1}_{ij}\;{}_jh_i \tag{15.39}$$

should be subtracted from the BCS (or self-consistent) energy in order to obtain the collective potential V_{coll}.

The generator coordinate method with the Gauss overlap approximation (GCM + GOA) has the advantage that the collective Hamiltonian (15.36) is not obtained by somewhat arbitrary quantization rules. It also clearly defines the form of Hamiltonian's kinetic and potential collective parts. In the Pauli and Podolsky prescription (13.1 - 13.4) of quantization, the potential part of the quantum Hamiltonian is simply indefinite.

It should be added that next to the Gaussian overlap approximation discussed above, there are other methods to solve the variational GCM problem (15.10), which also lead to collective Hamiltonians. One can recall here, e.g., the Symmetric Moment Expansion (SME) method, which gives results very close to those by GCM+GOA. In the last years, thanks to more advanced computers, the variational problem (15.10) may also be solved numerically.

The freedom of choosing the generating function in the GCM method makes it possible to describe dynamic phenomena using different coordinate systems, not necessarily position and momentum, as one does in quantum mechanics. It makes the currently used quantum mechanics one of the possible ways of describing the micro-world. The possibilities of the theory going beyond the traditional human view, based on the coordinates used so far, make the GCM method a universal way of describing the world.

Bibliography

[1] A. Bohr, B. Mottelson, *Nuclear Structure*, vol. I and II, World Scientific, Singapore, 1998 (reprints 1999, 2008).

[2] G.E. Brown, *Uniform Theory of Nuclear Models and Forces*, North-Holland, Amsterdam, 1967.

[3] A. Das, T. Ferbel, *Introduction to Nuclear and Particle Physics*, World Scientific, Singapore, 2003.

[4] J. Eisenberg, W. Greiner, *Nuclear Theory: Nuclear Models*, North-Holland, Amsterdam, 1987.

[5] R.W. Hasse, W. D. Myers, *Geometrical Relationships of Macroscopic Nuclear Physics*, Springer Verlag, Berlin - Heidelberg, 1988.

[6] K. Heyde, *The Nuclear Shell Model*, Springer Verlag, Heidelberg, 1994.

[7] H.J. Krappe, K. Pomorski, *Theory of Nuclear Fission*, Lecture Notes in Physics, vol. 838, Springer Verlag, Heidelberg, 2013.

[8] S.G. Nilsson, I. Ragnarsson, *Shapes and Shells in Nuclear Structure*, Cambridge University Press, Cambridge, 2005.

[9] S. C. Pancholi, *Pear-Shaped Nuclei*, World Scientific, Singapore, 2020.

[10] M. A. Preston, R. K. Bhaduri, *Structure of the Nucleus*, Addison-Wesley, Reading Massachusetts, 1975.

[11] P. Ring, P. Schuck, *The Nuclear Many Body Problem*, Springer Verlag, New York, 1980 (reprint 2005).

[12] D.J. Rowe, *Nuclear Collective Motion, Models and Theory*, World Scientific, Singapore, 2010.

[13] D.J. Rowe, J.L. Wood, *Fundamentals of Nuclear Models*, World Scientific, Singapore, 2010.

[14] R. Sachs, *Nuclear Theory*, Addison-Wesley, Cambridge, Mass. 1953.

[15] P.J. Siemens, A.J. Jensen, *Elements of Nuclei*, Addison-Wesley, Reading Massachusetts, 1987.

[16] J.D. Walecka, *Theoretical Nuclear and Subnuclear Physics*, World Scientific, Singapore, 2004.

Index

FSC
www.fsc.org
MIX
Papier | Fördert gute Waldnutzung
FSC® C083411

Kolibri 360 GmbH
Ferdinand-Jühlke-Straße 7
99095 Erfurt, Deutschland
produktsicherheit@kolibri360.de